Data Communications and Networks:
An Engineering Approach

Data Communications and Networks: An Engineering Approach

James Irvine and David Harle

Both of
University of Strathclyde, Glasgow, UK

JOHN WILEY & SONS, LTD

Copyright © 2002 by John Wiley & Sons, Ltd
Baffins Lane, Chichester
West Sussex, PO19 1UD, England

National 01243 779777
International (+44) 1243 779777
e-mail (for orders and customer service enquiries): cs-books@wiley.co.uk

Visit our Home Page on http://www.wiley.co.uk or http://www.wiley.com

Other Wiley Editorial Offices

John Wiley & Sons, Inc., 605 Third Avenue,
New York, NY 10158-0012, USA

WILEY-VCH Verlag GmbH
Pappelallee 3, D-69469 Weinheim, Germany

John Wiley & Sons Australia Ltd, 33 Park Road, Milton,
Queensland 4064, Australia

John Wiley & Sons (Canada) Ltd, 22 Worcester Road
Rexdale, Ontario, M9W 1L1, Canada

John Wiley & Sons (Asia) Pte Ltd, 2 Clementi Loop #02-01,
Jin Xing Distripark, Singapore 129809

British Library Cataloguing in Publication Data

A catalogue record for this book is available from the British Library

ISBN 0471 80872 5

Produced from Postscript files supplied by the authors.
Printed and bound in Great Britain by Antony Rowe Ltd, Chippenham, Wiltshire.
This book is printed on acid-free paper responsibly manufactured from sustainable forestry, in which at least two trees are planted for each one used for paper production.

Contents

Abbreviations

AAL	ATM Adaption Layer
ABR	Available Bit Rate
ACELP	Algebraic CELP
ADC	Analogue to Digital Conversion
ADCCP	Advanced Data Communications Protocol
ADPCM	Adaptive Differential Pulse Code Modulation
ADSL	Asymmetric Digital Subscriber Line
AFC	Access Control Field
AI	Air Interface
AM	Amplitude Modulation
AMI	Alternative Mark Inversion
AMPS	Advanced Mobile Phone System
ANSI	American National Standards Institute
APK	Amplitude Phase Keying
ARP	Address Resolution Protocol
ARPA	Advance Research Projects Agency
ARQ	Automatic Repeat reQuest
ASCI	Advanced Speech Call Items (GSM)
ASK	Amplitude Shift Keying
AT	ATtention
ATD	Asynchronous Time Division multiplexing
ATM	Asynchronous Transfer Mode
AuC	Authentication Centre (also AC)
BEB	Binary Exponential Back-off
BER	Bit Error Rate
BGF	Border Gateway Protocol
B-ISDN	Broad Band – Integrated Services Digital Network
BnZS	Bipolar with n zeros substituted
BOOTP	Bootstrap Protocol
BPSK	Binary Phase Shift Keying
BS	Base Station
BSC	Binary Symmetric Channel; Base Station Controller
BSI	British Standards Institute
BSS	Base Station Subsystem

BTS	Base Transceiver Station
BW	Bandwidth
CAC	Call Admission Control; Connection Admission Control
CAP	Carrierless Amplitude and Phase
CATV	Cable Access Television
CBC	Cipher Block Chaining
CBR	Constant Bit Rate Service
CC	Call Control
CCITT	Comité Consultatif International Télégraphique et Téléphonique (International Telegraph and Telephone Consultative Committee)
CCR	Commitment, Concurrency and Recovery
CDMA	Code Division Multiple Access
CDPD	Cellular Digital Packet Data
CDV	Cell Delay Variation
CELP	Codebook-Excited Linear Predictive
CEPT	Conférence Européenne des Postes et Télécommunication
CER	Cell Error Rate
CGI	Common Gateway Interface
CIDR	Classless Inter-Domain Routing
CIR	Carrier to Interference Ratio
CL	Link Connectivity
CLR	Cell Loss Rate
CMI	Coded Mark Inversion
CMIP	Common Management Information Protocol
CMIS	Common Management Information System
Cn	Node Connectivity
CNLS	connectionless (packet data service)
CNM	Central Network Management
CONP	Connection-Oriented Network Protocol
CPCS	Common Part Convergence Sub-layer
CRC	Cyclic Redundancy Check
CSMA/CA	Carrier Sense Multiple Access / Collision Avoidance
CSMA/CD	Carrier Sense Multiple Access / Collision Detection
CTD	Cell Transfer Delay
CVDT	Cell Variation Delay Tolerance
DARPA	Defence Advance Research Projects Agency
DAWS	Digital Advanced Wireless System
dB	decibel
dBm	decibel with reference to a milli-watt power
DCE	Data Communication Equipment
DDB	Distributed Database
DECT	Digital Enhanced Cordless Telecommunication
DES	Data Encryption Standard
DIFS	Distributed Interface Specification; Distributed Inter Frame Space
DMSP	Distributed Mail System Protocol

DMT	Discrete Multi-Tone
DNS	Domain Name System
DPP	Demand Priority Protocol
DPSK	Differential Phase Shift Keying
DQDB	Distributed Queue Dual Bus
DQPSK	Differential Quaternary Phase Shift Keying
DSB-AM	Double Sideband Amplitude Modulation
DSCP	Diff Serv Control Protocol
DSL	Digital Subscriber Line
DSRR	Digital Short Range Radio system
DSS	Digital Signature Standard
DTE	Data Terminal Equipment; Data Termination Equipment
DTMF	Dual Tone Multi Frequency
DVMRP	Distance Vector Multicast Routing Protocol
ECMA	European Computer Manufacturers Association
EDACS	Enhanced Digital Access Communication System
EGP	Exterior Gateway Protocol
EIA	Electrical Industries Association
EIRP	Effective Isotropic Radiated Power
EMC	ElectroMagnetic Compatibility
ER	Explicit Rate
ETSI	European Telecommunications Standards Institute
FCC	Federal Communications Commission
FCS	Frame Check Sequence
FDDI	Fibre Distributed Data Interface
FDM	Frequency Division Multiplexing
FDMA	Frequency Division Multiple Access
FEC	Forward Error Correction
FHMA	Frequency Hopping Multiple Access
FM	Frequency Modulation
FSK	Frequency Shift Keying
FTP	File Transfer Protocol
GCRA	Generic Cell Rate Algorithm
GMSK	Gaussian Minimum Shift Keying
GoS	Grade of Service
GPRS	General Packet Radio Service
GSM	Global System for Mobile communications
GSM-R	GSM for Railways
HDB	Home DataBase
HDB3	High Density Bipolar
HDLC	High Level Data Link Control
HDTV	High Definition Television
HLR	Home Location Register

HSCSD	High Speed Circuit Switched Data
HSLN	High Speed Local Network
HTML	Hyper Text Mark Up Language
HTTP	Hyper Text Transfer Protocol
ICMP	Internet Control Message Protocol
ICP	Initial Connection Protocol
IDC	Insulation Displacement Connector
IDEA	International Data Encryption Algorithm
iDEN	Integrated Digital Enhanced Technology
IDFT	Inverse Discrete Fourier Transform
IDU	Interface Data Unit
IEEE	Institute of Electrical and Electronics Engineers
IF	Intermediate Frequency
I-MAC	Isochronous MAC
IMAP	Interactive Mail Access Protocol
IMEI	International Mobile Equipment Identity
IMP	Interface Message Processor
IMSI	International Mobile Subscriber Identity
IMTS	Improved Mobile Telephone Service
IN	Intelligent Network
IP	Internet Protocol
IPX	Internetwork Packet Exchange
ISDN	Integrated Services Digital Network
ISI	Inter-Symbol Interface
ISO	International Standards Organisation
ISP	Internet Service Provider
ITU	International Telecommunications Union
ITU-R	Radio communications sector of the ITU
ITU-T	Telecommunications standardisation sector of the ITU
IV	Initial Value
JPEG	Joint Photographic Experts Group
KSG	Key Stream Generator
LA	Location Area
LAC	Location Area Code
LAN	Local Area Network
LAP	Link Access Procedure
LAP-B	Link Access Protocol – B
LAP-D	Link Access Protocol – D
LCN	Logical Channel Number
LCP	Link Control Protocol
LED	Light Emitting Diode
LFSR	Linear Feedback Shift Register
LIS	Logical IP Subnet

LLC	Logical Link Control
LMN	Land Mobile Network
LMR	Land Mobile Radio
LNM	Local Network Management
LPC	Linear Predictive Coding
LS	Line(-connected) Station
LSR	Label Switched Routers
LTR	Logic Trunked Radio
LTU	Line Termination Unit
MAC	Media Access Control
MAN	Metropolitan Area Network
MCC	Mobile Country Code
MCR	Minimum Cell Rate
MEGACO	Media Gateway Control Protocol
MER	Message Error Rate
MIB	Management Information Base
MIME	Multipurpose Internet Mail Extension
MLT-3	Multi-Level Transition 3
MMF	Multi-Mode Fibre
MMI	Man Machine Interface
MNC	Mobile Network Code
MNI	Mobile Network Identity
MoU	Memorandum of Understanding
MPEG	Motion Picture Experts Group
MPT	Ministry of Post and Telecommunications
MS	Mobile Station
MSC	Mobile Switching Centre
MSK	Minimum Shift Keying
MTBF	Mean Time Between Failure
MTP	Message Transfer Part
MTTR	Mean Time to Repair
NAK	Negative Acknowledgement
NAP	Network Access Point
NCP	Network Control Protocol (related to PPP);
	Network Core Protocol (Novell Netware)
NIC	Network Information Centre
N-ISDN	Narrow Band – Integrated Services Digital Network
NIST	National Institute of Standards and Technology
NMU	Network Management Unit
NNI	Network Network Interface
NPA	Network Point of Attachment
NRZ	Non Return to Zero
NRZI	Non Return to Zero Invert to one
NSAP	Network Service Access Point
NUA	Network User Address

OMC	Operations & Management Centre
OOK	On-Off Keying
OQPSK	Offset Quaternary Phase Shift Keying
OSI	Open Systems Interconnection
OSPF	Open Shortest Path First
OTAR	Over The Air Re-keying

PA	Pre-Arbitrated Access; Power Amplifier
PAD	Packet Assembler/Disassembler
PAM	Pulse Amplitude Modulation
PAMR	Public Access Mobile Radio
PBR	Private Business Radio
PBX	Private Branch Exchange
PC	Personal Computer; Protocol Control
PCM	Pulse Code Modulation
PCR	Peak Cell Rate
PD	Packet Data
PDA	Personal Digital Assistant
PDN	Public Data Network
PDU	Protocol Data Unit
PEI	Peripheral Equipment Interface
PICS	Protocol Implementation Conformance Statement
PIN	Personal Identity Number
PINX	Private Integrated Network eXchange
PISN	Private Integrated Services Network
PLMN	Public Land Mobile Network
PLP	Packet Level Protocol
PM	Phase Modulation
PMD	Physical Media Dependent
PMR	Private Mobile Radio
PN	Pseudo raNdom
POTS	Plain Old Telephone System
PPP	Point to Point Protocol
PSK	Phase Shift Keying
PSS	Packet Switched Service
PSTN	Public Switched Telephone Network
PTT	Press To Talk; Post, Telegraphy & Telecommunications
PVC	Permanent Virtual Circuit; Permanent Virtual Connection

QA	Queue Arbitrated Access
QAM	Quadrature Amplitude Modulation
QoS	Quality of Service
QPSK	Quadrature Phase Shift Keying

| RARP | Reverse Address Resolution Protocol |
| RES | Radio Equipment and System |

RFC	Request For Comment
RF	Radio Frequency
RIP	Routing Information Protocol
RL	Ring Latency
ROSE	Remote Operations Service Element
RPE	Regular Pulse Excitation (compression technique)
RSA	Rivest Shamir Adleman (cryptographic system)
RSSI	Received Signal Strength Indicator
RSVP	Resource Reservation Protocol
RTCP	Real Time Control Protocol
RTP	Real Time Protocol

SAGE	Security Algorithms Group of Experts
SAP	Service Access Point; Service Advertising Protocol (Novell Netware)
SC	Sub-Committee
SCR	Sustainable / Sustained Cell Rate
SDH	Synchronous Digital Hierarchy
SDL	Simple Data Link
SDLC	Synchronous Data Link Control
SDS	Short Data Service
SDU	Service Data Unit
SIFS	Short Interframe Space
SIM	Subscriber Identity Module
SIP	Session Initiation Protocol
SIR	Signal to Interference Ratio
SLA	Service Level Agreement
SLIP	Serial Line IP
SMDS	Switched Multi Megabit Data Service
SMF	Single Mode Fibre
SMR	Specialist Mobile Radio
SMS	Short Message Service
SMT	Station Management
SMTP	Simple Mail Transfer Protocol
SNMP	Simple Network Management Protocol
SNR	Signal to Noise Ratio
SONET	Synchronous Optical Network
SPX	Sequenced Packet Exchange
SRBR	Short Range Business Radio
SS7	Signalling System 7
SSI	Short Subscriber Identity
STD	Synchronous Time Division multiplexing
STE	Signalling Terminal Exchange
STP	Shielded Twisted Pair; Spanning Tree Protocol
SVC	Switched Virtual Circuit
SwMI	Switching and Management Infrastructure

TA	Terminal Adapter (or Adapting)

TAC   Type Approval Code
TACS   Total Access Communication System
TC   Technical Committee
TCM   Trellis Code Modulation
TCP   Transmission Control Protocol
TDD   Time Division Duplex
TDM   Time Division Multiplexing
TDMA   Time Division Multiple Access
TE   Terminal Equipment
TETRA   TErrestrial Trunked RAdio (Private Mobile Radio system)
TMN   Telecommunications Management Network
TSAP   Transport Service Access Point
TTRP   Target Token Rotation Protocol
TTRT   Target Token Rotation Time

UBR   Unspecified Bit Rate
UDP   User Datagram Protocol
UNI   User Network Interface
UPC   Usage Parameter Control
URI   Universal Resource Identifier
URL   Uniform Resource Locator
UTP   Unshielded Twisted Pair

VAD   Voice Activity Detector
VBR   Variable Bit Rate
VC   Virtual Circuit; Virtual Channel
VCI   Virtual Channel Indicator
VLAN   Virtual LAN
VoD   Video on Demand
VoIP   Voice over IP
VP   Virtual Path
VPI   Virtual Path Indicator
VPN   Virtual Private Network
VSELP   Vector Sum Excited Linear Predictive

WAN   Wide Area Network
WAP   Wireless Application Protocol
WDM   Wavelength Division Multiplexing
WG   Working Group
WLAN   Wireless Local Area Network
WLL   Wireless Local Loop
WT   Waiting Time
WWW   World Wide Web
WYSIWYG   What You See Is What You Get

Preface

It was once suggested that the transfer of information from one user to another is rather like a swan swimming from one side of a pond to another. At the top level, it appears as effortless elegance and simplicity, whilst underneath the water, invisible to all, there is significant activity, energy and not a little chaos. We are familiar now, contemptuous even, with our ability to transfer digital information over communication networks. Witness the exponential growth of the Internet, the widespread adoption of digital mobile telephone and its associated SMS capabilities. We almost take the universal service provision of communication services for granted; expectation could be said to have now triumphed over hope. However, such apparent ease of service provision and use belies the complexity of the communication techniques that underpin such ubiquitous and widespread services. In this book, we consider the goals of data communications first from the service perspective, examining in detail the effect that user demands have upon network design. The top-down approach is continued by considering the principles and technologies associated with such network provision, followed by link issues and completed by a consideration of the key principles associated with the transfer of individual bits. The book rounds off by considering case studies that illustrate how such concepts are combined in practice to achieve a seamless service that meets the needs of the users: the engineering approach. This contrasts with other texts where either a bottom-up approach is adopted (signals up to services) or where a service perspective is taken but is limited to a packet view and stops short of considering the physical layers.

Readership

Data Communications and Networks: An Engineering Approach is a book that is intended to be read, rather than act as a reference book, kept on a shelf and visited only whenever a particular piece of information is sought. After reading this book, we hope that the reader will have gained an understanding of the key concepts associated with the transfer of information over communication networks. Such an understanding will, in turn, enable the reader to effectively distil and use the information from the myriad of books, standards, online material, articles, tutorials and other sources now available.

Data Communications and Networks: An Engineering Approach was written for anyone, students or professionals with a Science or Engineering background, who wish to gain a basic understanding of the key concepts behind data communications engineering. Examples of such readership groups are undergraduate students taking penultimate and final year classes in Data Communications, postgraduate students taking 1st semester or conversion modules, or Engineering professionals who wish to make their initial foray into data communications or require an update on the basic principles. Although not overtly mathematical in nature, the book uses both mathematical and statistical techniques to describe Data Communications and Network concepts. Consequently, we have assumed a mathematics and statistics knowledge equivalent to that gained after two or three years of a science or engineering.

Support

In order to provide further insight into the areas covered in the book, a web site with additional examples and links to other material has been established at `http://comms.eee.strath.ac.uk/datacomms`. The web site also has Java applets to illustrate key points and algorithms. Please feel free to visit and to send us your comments.

Acknowledgements

In the course of writing and producing this book, the authors are indebted to a great many people who have made contributions both directly and indirectly. Firstly we would like to acknowledge the efforts of all the staff at Wiley and, in particular, Mark Hammond and Sarah Hinton who showed both faith and a great deal of patience to enable us to complete this book. Additionally, thanks are due to our colleagues within the Department of Electronic and Electrical Engineering who supported, commented upon and refined aspects of the material contained within. Our discussions on technological and pedagogical issues were always lively, stimulating and informative; some of which we hope is reflected in this text. Whether they realised at the time or not, a contribution has also been made by our students who acted as guinea pigs for much of the material and questions and whose only potential rewards were the credits at the end of the year. Finally, the authors would like to express heartfelt thanks to our families for all their support throughout this extra-mural endeavour; without whom completion would not have been possible. Cheers.

1

Communication Systems

1.1   Introduction

This book provides a technical introduction to the general issues associated with data communications. Data communications is the problem of getting *information* from one place to another *reliably* (secure both from channel disruptions and deliberate interference) while conforming to user *requirements*. The key points – the information itself, its reliable transmission, and conformance to requirements – are dealt with in this book.

Communications have increased significantly in importance in the last few years. It is often said that information is power, but in fact that power derives from having the information that you need when and where you need it. We are very much more mobile than our forefathers, and as we move about more, and trade over larger distances, the need for communications increases. On the other hand, the easier communication becomes, and the more available it is, the more people want to use it.

Successful communication systems are built on successful applications. The user is literally the beginning and end of the system. This has been shown recently in the differing success of voice services and WAP (Wireless Application Protocol) services for mobile phones. Voice services have seen an unprecedented increase in use with more than half the population of the UK now having a mobile phone. Mobile phones are easy to use, convenient, and relatively inexpensive. A voice conversation can be carried out almost anywhere. On the other hand, WAP services are difficult to use, expensive for the information obtained, and there are easier ways of obtaining the same result. This may well change if more people use PDAs to access information on the move, but WAP on mobile phones has been very much less successful than hoped.

1.2   Partitioning Communication Systems

This book starts from the point of view of the user rather than from the problem of the communication at a low level. Users do not interact directly with the communication system, but interact with applications which themselves communicate (Figure 1.1).

For example, a user may well use the Web browser on their computer to interact with a Web server on a remote computer. However, these applications use the underlying communication system in order to communicate. It would be very restrictive for the communication system to handle each application individually. This would mean that whenever a new application was developed, the communication system would have to be changed. A more successful way is for the application to format its requests from a fixed set of *services* provided by the

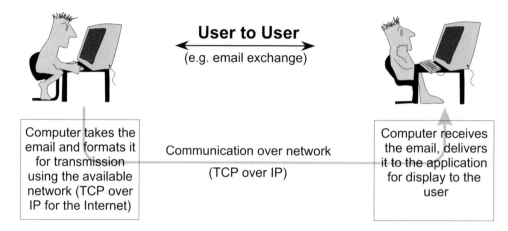

Figure 1.1   The communications system from the user's viewpoint

communication system. A good example of this is a fax machine, which translates documents over a communication system that was designed for the application of voice calls. As fax machines can make use of the existing infrastructure, they could be installed with little cost or inconvenience. This meant that a large installed base of fax machines could be developed very quickly, which meant in turn that more people were encouraged to get fax machines, and so on.

For this reason, most communication systems provide a known set of interfaces to a set of services. Applications then convert what their communication requirements into these services. This has the advantage that the application does not need to know anything about the details of the communication service system, simply about the service interface to which it directs its data. The communication system, on the other hand, does not have to worry about the details of the application. All it has to do is carry the advertised service from the source to the destination. A final advantage is that it may be possible for two different applications to communicate if they both use the communication services in the same way. For example, a Microsoft Outlook user can email a Netscape Communicator user, or a dial telephone user can call a push button phone user.

A telephone system is designed to send a varying electrical signal from one subscriber to another subscriber. The application in this case is the phone itself, and the service interface is a telephone plug in the wall. You can plug whichever phone you like into the socket as long as a phone can understand the basic interface that the network provides, in this case the varying electrical signal for speech and for dialling. If you plug a fax machine into the wall, your application will change, but no change to the communication system is required because the communication service is the same.

We can continue this approach to divide up the rest of the communication system. To provide a direct communication link from every source to every destination would be impractical. To avoid this problem, all but the very simplest of communication systems will consist of a network, a series of switches and links to connect the source and the destination.

The lowest level in this hierarchy is the communication over an individual link. This involves getting data from one place to another over a given transmission medium.

This four-level hierarchy, with applications making use of services, switched over a network of links, each of which transports data from place to place over a transmission medium, is called a layered hierarchy because at any level we can abstract the layers below.

1.3 Layered Communication Architectures

The layered structure can be defined more formally. Communication systems are partitioned into a vertical set of layers, with each layer performing a related subset of the functions required to communicate with another system. The functions within a layer are collected into groups called *entities*, and it is these entities within the same layer on different systems which communicate with their peers using one or more *protocols*, defined rules for transferring information or instructions. The architecture is therefore often referred to as a *protocol stack*. Any particular layer will depend upon the layer below (if there is one) to perform more primitive functions and to conceal the details of these functions. The interfaces between the layers are known as *Service Access Points* (SAP) (see Figure 1.2).

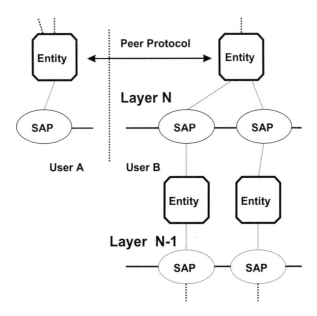

Figure 1.2 Service access points

Communicating entities in a given layer exchange messages called Protocol Data Units (PDUs). The protocol syntax defines the format and the semantics defines the meaning of the messages. The syntax must be precisely defined to avoid confusion. The PDU for a layer is made up of protocol information for that layer along with the PDU from the layer above as payload.

Entities use peer protocols to communicate with each other in order to implement their service definition. The protocol used can be changed provided that the service visible to the service user does not change. The service and protocol are always decoupled.

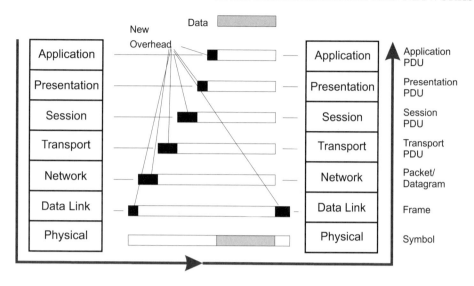

Figure 1.3   PDUs for various layers

1.3.1   The Need for Standards

As noted above, the protocol must be agreed by the communicating entities in order for the system to work. Since these entities may even be in different countries, there is obviously a need for standards to be agreed. Standards can be *de facto*, 'from the fact', such as UNIX or the IBM PC, or *de jure*, 'by law', formal legal standards adopted by an authorised standards body. These may be national, like the ANSI, BSI or ARIB, or international, governed by treaty among national governments, like the International Telecommunications Union (ITU).

Standards are arrived at by a variety of means and are produced by a variety of organisations. These include government bodies, intergovernmental organisations, professional bodies and industry groups.

The International Standards Organisation (ISO) is a voluntary non-treaty organisation and issues standards on a wide range of subjects. It has nearly 90 members including ANSI, BSI, AFNOR, and DIN and consists of around 200 technical committees (TCs) numbered in order of creation dealing with topics a diverse as nuts and bolts (TC1) and computers and information processing (TC97). Each TC is made up of Sub-committees (SGs) which are further divided into Working Groups (WGs). The work is carried out mainly by WGs that comprise volunteers from academics, government officials and industrial representatives. The ISO cooperates with the ITU to avoid incompatible standards. The ITU was formally the CCITT, and consists of the ITU-R, the radiocommunications sector, and the ITU-T, the telecommunication standards sector, along with a secretariat and a development sector.

It in interesting to note a significant difference between the approaches taken by the United States and Europe on the issue of standards. Europe takes a very proactive stance on the subject of standards. ETSI, the European Telecommunications Standards Institute, has a very active programme of standardisation and manufacturers of communication equipment are often required by European governments to produce equipment which conforms to these standards if it is allowed to be sold in the European Union. The situation in the United States of America is different with a much more 'hands-off' approach, letting the market decide. One of the major bodies producing standards there is the Institute of Electrical and Electronics Engineers (IEEE), which is a professional body and not part of the government.

1.3.2   Internet Reference Model

The four-level model is one of the simplest that can be applied to communication systems, but there are others. While there is no full standard Internet protocol stack, the architecture, which was developed from an original ARPANET network, has five layers: Application, Transport, Internet, Data Link and Physical (Figure 1.4).

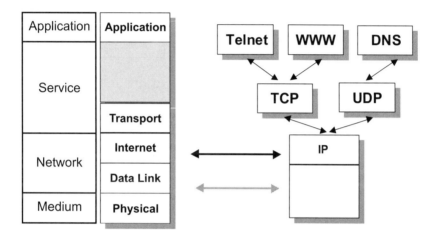

Figure 1.4   The ARPANET (Internet) protocol stack

The difference between the simple four-layer architecture and the Internet architecture is that the network layer is split in two between a 'network' layer for an individual communications network (termed the Data Link Layer), and the Internet layer which acts as a network of networks.

1.3.3   OSI Reference Model

The Internet model has its basis in computing. One of the most common telecommunications models is the OSI (Open Systems Interconnection) model, which has seven layers, as shown in Figure 1.5. It is often referred to as the seven-layer model.

The thought behind the Open Systems Interconnection model was the advantage of interoperability mentioned above, i.e., if the model is used by all manufacturers, any machine which implements the OSI model will be 'open', i.e. able to connect to, all other systems following the OSI model throughout the world. Developed in 1979 by the ISO, it aimed to provide a common basis for the co-ordination of standards development for system interconnection, while allowing existing standards to be put into perspective within the overall reference model.

The layers are numbered from 1 at the bottom to 7 at the top, and assume that in the diagram there is a user A at the top left-hand side wishing to communicate with user B at the top of the right-hand stack. The information from A has to be sent down the layers to layer 1, where it

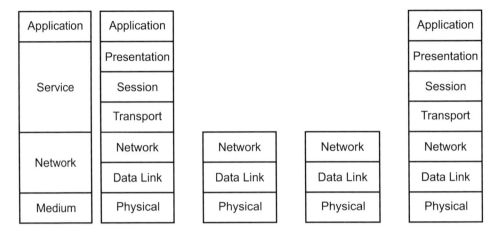

Figure 1.5   The OSI protocol stack

is transmitted across the network to layer 1 associated with user B; it then travels up through the layers to user B. In Figure 1.5, an intermediate node is shown between the two nodes. This node is concerned with routing the information from user A to user B (and the reverse direction) and as such only performs functions associated with the first three layers; hence it is indicated by the small stack at the centre. This is often the case in a network, with the four upper layers only being present in the end points. An exception is if transcoding from one type of data to another is required, when the remaining layers would be used.

The layers of the model can be divided into three categories. The top four layer protocols have the functions of organising and presenting the data in a manner that makes sense eventually to the end user. These correspond to the service and application layers. Protocols in these layers must be consistent for pairs of communicating end user programs.

Layer 3 is concerned with how to route messages between applications in the presence of intermediate nodes. Protocols at this level must be consistent throughout a given network.

The bottom two layers are concerned with physical communication of messages between adjacent nodes. Protocols must be consistent between each pair of adjacent nodes.

- **Application (Layer 7)**: The application layer provides the user interface to the communications system. Examples include electronic mail, distributed data bases and network operating systems.
- **Presentation**: The presentation layer provides a common format to the application layer. It allows hosts to communicate even if they use different representations. For example, different web browsers provide different user interfaces – different applications – but all understand the HTML page format, the JPEG images format, etc. These file formats are examples of presentation layer protocols, while the browsers themselves reside in the application layer. Services provided by the presentation layer therefore include things like file transfer, virtual terminal protocols, compression, code transformation and encryption.
- **Session**: The session layer breaks down the services sent from the presentation layers into basic data flows which can be sent over the system. It manages interactions between pairs of communicating application processes. 'Sessions' can be short or long, involving one or many messages interactions e.g. bank cash dispenser to computer; terminal logged into remote computer. Services provided to the presentation layer include session

establishment/termination, authentication/accounting, synchronisation, data transfer and exception reporting.

- **Transport**: The transport layer provides the initial establishment of a communications channel, the transfer of the data and the final release of the channel. For data transfer, it interfaces with the network layer to ensure an error-free virtual point-to-point connection, where the communications arrive in the correct order. In the case when computers have multi-tasking capabilities, the transport layer ensures that the communication goes to the correct process and performs any multiplexing or demultiplexing required.
- **Network**: The network layer interconnects data link communications paths into a global network that connects all open systems, i.e. the routing of the communications from the source node to the destination node. It provides the rules for routing packets through the network, and for flow control where necessary, to ensure that not too many packets are allowed in the system.
- **Data Link**: This layer ensures reliable communications over the physical layer. The protocol determines the structure of data, i.e. frame or packet size, and deals with aspects such as flow control, error detection and error recovery.
- **Physical**: This layer is concerned with the physical aspects of the communications link – mechanical and electrical – and provides the means to transmit bits of data across a continuous communications path. The protocol for this layer sets details such as cable size, loss and frequency characteristics, connector types, pin arrangements, voltage levels and transmission coding.

The OSI model clearly defines Services, Interfaces and Protocols. The Internet architecture is very simpler. It was developed from a telecommunications perspective whereas the Internet model is really a computer-based idea. It has a large number of layers which do not always have a well-defined role. Others, like the data link layer, are unwieldy and are often split into sub-layers. It was designed to address the general problem of communication and was developed before the protocols. Lack of experience with protocols and dealing with vested interests resulted in an over-complex solution. On the other hand, the Internet model is not actually a formal model but is simply fitted around working protocols. It has fewer layers but is not good at describing things other than TCP/IP. The OSI model itself is popular, but its protocols are not so widely used. TCP/IP is more popular.

1.4   Structure of the Book

This book follows the layering model, although unlike most books, we start from the upper layers and work downwards (see Figure 1.6).

The layering model is a bit abstract, so the best way to see what is happening is to consider an example. The Internet provides an excellent example of a communications system, particularly as there are various tools available which can be used to see what is actually going on.

To show the various principles in operation, take the case of looking at the home page of the Institute for Telecommunication Research at the University of South Australia. The home page itself is shown in Figure 1.7. This is how it would appear to the user in one type of browser. A different browser may display it slightly differently. The web browser itself is a layer 7 application.

The web browser is interpreting a HTML page, which is an example of a layer 6 protocol.

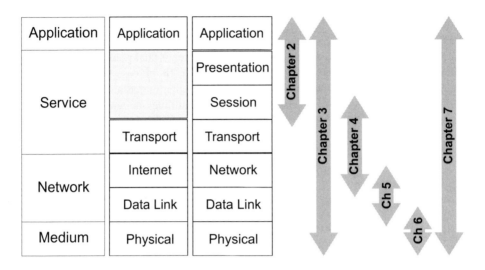

Figure 1.6   Structure of the book

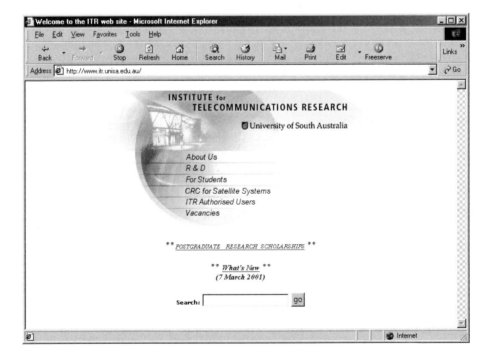

Figure 1.7   Example home page

The web page which looks like Figure 1.7 is encoded as follows:

```
<!DOCTYPE HTML PUBLIC "-//W3C//DTD HTML 3.2 Final//EN">
<html>
<head>
<title>Welcome to the ITR web site</title>
</head>
<body link="#0000FF" alink="#800080" vlink="#FF0000" text="#000000"
leftmargin="0" topmargin="0" bgcolor="#FFFFFF">
...
</body>
</html>
```

Since this example uses the Internet, there is no well-defined session layer. However, the interaction of the application with the network routines within the computer operates in this way. This interaction is shown below. First, our computer connects to the ITR computer and requests the file http://www.itr.unisa.edu.au/. It states that the protocol it is using is HTTP/1.1, that it can accept anything, that it would like English, what type of browser it is, and so on. The ITR computer responds by sending a result code of 200 to say that the request is okay and the file will follow. It then sends back the file. However, during the receipt of the file, the browser notices that it contains an image, and opens a second connection to get the image. Since this is a rather slow connection, the data is delivered to the application in chucks. These chunks are generally multiples of 536 bytes, which is the size of TCP segments used here.

CONNECTION 1: Connecting to server
CONNECTION 1: Client to Server (250 bytes)

```
GET http://www.itr.unisa.edu.au/ HTTP/1.1
Accept: */*
Accept-Language: en-gb
Accept-Encoding: gzip, deflate
User-Agent: Mozilla/4.0 (compatible; MSIE 5.5; Windows 98)
Host: www.itr.unisa.edu.au
Proxy-Connection: Keep-Alive\
Pragma: no-cache
```

CONNECTION 1: Server to Client (536 bytes)

```
HTTP/1.1 200 OK
Date: Sun, 29 Jul 2001 16:38:27 GMT
Server: Apache/1.3.9 (Unix)
Transfer-Encoding: chunked
Content-Type: text/html

<!DOCTYPE HTML PUBLIC "-//W3C//DTD HTML 3.2 Final//EN">
<html>
<head>
<title>Welcome to the ITR web site</title>
</head>
<body link="#0000FF" alink="#800080" vlink="#FF0000"
text="#000000" leftmargin="0" topmargin="0" bgcolor="#FFFFFF">
<div align="center"><center>
<table BORDER="0" CELLPADDING="0" CELLSPACING="0" WIDTH="100%" height="100%">
  <tr>
    <td valign="center" al
```

CONNECTION 1: Server to Client (1072 bytes)

```
align="center"><!--webbot bot="ImageMap" startspan
...
```

CONNECTION 1: Server to Client (1608 bytes)

```
HREF="tech_res/tech_res.html"><AREA SHAPE="RECT" COORDS="140, 106, 208, 124"
HREF="about_us/about.html"></MAP><img src="images/itr_new.gif" width
<font face="verdana, arial, helvetica" size="2"><form action="/cgi-bin/htsearch"
    method="post">
...
```

CONNECTION 2; Connecting to server
CONNECTION 2: Client to Server (307 bytes)

```
GET http://www.itr.unisa.edu.au/images/itr_new.gif HTTP/1.1
Accept: */*
Referer: http://www.itr.unisa.edu.au/
Accept-Language: en-gb
Accept-Encoding: gzip, deflate
User-Agent: Mozilla/4.0 (compatible; MSIE 5.5; Windows 98)
Host: www.itr.unisa.edu.au
Proxy-Connection: Keep-Alive
Pragma: no-cache
```

CONNECTION 1: Server to Client (450 bytes)

```
type="hidden" name="config" value="itr"><input type="hidden" name="method" value="or">
     <input type="hidden" name="format" value="builtin-long"><font face="verdana,
     arial, helvetica" size="1"><b><p>Search:</b> <input type="Text" name="words"
       size="20"> <input type="Submit" name="submit" value="go"> </font></font></p>
     </form>
     </td>
   </tr>
</table> </center></div><!--crc_sat_sys.html-->

</body>
</html>
```

CONNECTION 2: Server to Client (536 bytes)
HTTP/1.1 200 OK

. . .

CONNECTION 2: Server to Client (1072 bytes)
CONNECTION 2: Server to Client (1608 bytes)
CONNECTION 2: Server to Client (1608 bytes)
CONNECTION 2: Server to Client (1072 bytes)
CONNECTION 2: Server to Client (1608 bytes)
CONNECTION 2: Server to Client (1608 bytes)
CONNECTION 2: Server to Client (1072 bytes)
CONNECTION 2: Server to Client (536 bytes)
CONNECTION 2: Server to Client (536 bytes)
CONNECTION 2: Server to Client (8040 bytes)
CONNECTION 2: Server to Client (3752 bytes)
CONNECTION 2: Server to Client (4824 bytes)
CONNECTION 2: Server to Client (4824 bytes)
CONNECTION 2: Server to Client (4589 bytes)
CONNECTION 2: Server to Client (3216 bytes)
CONNECTION 2: Server to Client (3405 bytes)
CONNECTION 1: Server disconnected
CONNECTION 2: Server disconnected

The next layer down in the system is the transport layer. In TCP/IP, each TCP segment is carried on one IP packet, so the two are linked.

The activity at this level starts as shown in the following display. Each line gives an IP packet, and most packets contain TCP segments, although some contain UDP packets (see Section 4.12.3.1)for querying name servers. hestia.clara.net is the local name server which our computer (me) has been told to query for addresses. The first problem facing the computer is contacting the name server. In line 1, we issue an ARP packet requesting the address on the local network where the nameserver can be found. Having obtained this, we send a packet (line 2) from a connection on port 1300 to the name server asking for the address of http://www.itr.unisa.edu.au. The name server replies, saying that http://www.itr.unisa.edu.au is a canonical name, and that the real name is charli.levels.unisa.edu.au. Name server queries are covered in Section 4.13.5. We then send another ARP query to find out the node to which we send requests for charli.levels.unisa.edu.au. Of course, charli.levels.unisa.edu.au is not on our network, but the router on our network responsible for taking traffic to other networks will respond with its address. Up to this point, all the activity has just been to set up the data connection, rather than send any data itself. Having the necessary address, we send it as a segment to port 80, the web server port, requesting that a connection be set up (line 5). The web server responds (line 6), acknowledging the request and granting the connection, and we acknowledge this (line 7). This is the three-way handshake to set up a TCP connection. Finally, at line 8, we send the data which contains the request we covered in the previous section. The server responds (line 9) acknowledging, but not yet sending any data. The data follows in line 10, which we acknowledge in line 11. This connection has negotiated a maximum segment size of 536 bytes, so TCP segments have this size. Two more follow from the server, both of which are acknowledged (line 14). More data follows in segments in lines 15, 16, 18 and finally 21, being acknowledged in lines 17 (two segments), 19 and 22. Lines 20, 23 and 24 are the three-way handshake for the second connection to get the image.

```
 1   arp request for hestia.clara.net
 2   me.1300 > hestia.clara.net.53:              1+ A? www.itr.unisa.edu.au. (38)
 3   hestia.clara.net.53 > me.1300:              1 2/3/3 CNAME cha (201)
 4   arp request for charli.levels.unisa.edu.au
 5   me.1305 > charli.levels.unisa.edu.au.80:    S 5637114:5637114(0) win 8192
                                                 <mss 536,sackOK>
 6   charli.levels.unisa.edu.au.80 > me.1305:    S 2454387047:2454387047(0) ack 5637115
                                                 win 9112 <mss 536>
 7   me.1305 > charli.levels.unisa.edu.au.80:    . ack 1 win 8576
 8   me.1305 > charli.levels.unisa.edu.au.80:    P 1:286(285) ack 1 win 8576
 9   charli.levels.unisa.edu.au.80 > me.1305:    . ack 286 win 9112
10   charli.levels.unisa.edu.au.80 > me.1305:    P 1:537(536) ack 286 win 9112
11   me.1305 > charli.levels.unisa.edu.au.80:    . ack 537 win 8576
12   charli.levels.unisa.edu.au.80 > me.1305:    . 537:1073(536) ack 286 win 9112
13   charli.levels.unisa.edu.au.80 > me.1305:    P 1073:1609(536) ack 286 win 9112
14   me.1305 > charli.levels.unisa.edu.au.80:    . ack 1609 win 8576
15   charli.levels.unisa.edu.au.80 > me.1305:    . 1609:2145(536) ack 286 win 9112
16   charli.levels.unisa.edu.au.80 > me.1305:    . 2145:2681(536) ack 286 win 9112
17   me.1305 > charli.levels.unisa.edu.au.80:    . ack 2681 win 8576
18   charli.levels.unisa.edu.au.80 > me.1305:    P 2681:3217(536) ack 286 win 9112
19   me.1305 > charli.levels.unisa.edu.au.80:    . ack 3217 win 8576
20   me.1308 > charli.levels.unisa.edu.au.80:    S 5641262:5641262(0) win 8192
                                                 <mss 536,sackOK>
21   charli.levels.unisa.edu.au.80 > me.1305:    P 3217:3667(450) ack 286 win 9112
22   me.1305 > charli.levels.unisa.edu.au.80:    . ack 3667 win 8126
23   charli.levels.unisa.edu.au.80 > me.1308:    S 2455265723:2455265723(0) ack 5641263
                                                 win 9112 <mss 536>
24   me.1308 > charli.levels.unisa.edu.au.80:    . ack 1 win 8576
```

At the link layer, the actual data sent to transport these IP packets will depend on the technology used. This varies throughout the journey. The trip to Adelaide has 23 hops. The start is at a laptop computer in Glasgow with a dial-up connection to the Claranet Internet Service Provider. The following list the route to the web server in Australia. The rather convoluted route involved is shown in Figure 1.8.

1. ftr-rk-35.access.clara.net(195.8.83.35) in London
2. access-fe-0-0-0-banner-starbuck.router.clara.net (195.8.83.128) in London (Claranet)

3. ge-5-0-0-banner-bildad.router.clara.net (195.8.68.62) in London (Claranet)
4. atm-6-0-0-telee-peleg.router.clara.net (195.8.68.158) in London
5. fastethernet1-0.lth-icr-02.carrier1.net in London (Carrier1 AG, Zurich)
6. fastethernet4-0-0.lth-bir-01.carrier1.net (212.4.203.209) in London
7. pos13-0.lon-bbr-02.carrier1.net (212.4.200.121) in London
8. pos13-0.nyc-bbr-02.carrier1.net (213.239.20.254)
9. gigabitethernet2-0.nyc-pni-03.carrier1.net (212.4.193.202)
10. p4-1.nycmny1-cr7.bbnplanet.net (4.24.163.17) in New York (BBN Planet)
11. p2-0.nycmny1-nbr1.bbnplanet.net (4.24.7.1) in New York
12. p1-0.nycmny1-br1.bbnplanet.net (4.24.10.82) in New York
13. p1-0.nycmny1-ba1.bbnplanet.net (4.24.6.230) in New York
14. p1-0.xnyc1-mci.bbnplanet.net (4.24.7.70) in New York
15. acr1-loopback.SantaClara.cw.net (208.172.146.61) in San Francisco (Cable & Wireless)
16. – (208.172.147.210)
17. optus-networks.Sydney.cw.net (166.63.225.166) in Sydney
18. GigEth12-0-0.ia4.optus.net.au (202.139.191.18) in Sydney (CWO Infrastructure Network)
19. SA-RNO-Int.ia4.optus.net.au (202.139.32.206) in Sydney
20. ethernet1.city-east.unisa.gw.saard.net (203.21.37.110) in Adelaide (South Australian Academic, Research and Development Network)
21. MLgate.levels.unisa.edu.au (130.220.10.2) Adelaide (Uni of South Australia)
22. MLCamp.levels.unisa.edu.au (130.220.2.106) in Adelaide
23. www.itr.unisa.edu.au (130.220.36.143) in Adelaide

Figure 1.8   Route to the ITR web server

The whole point of IP is to insulate the upper layers from the detail of the network, so it is not possible to tell what types of transmission are actually used, but the initial connection was over a telephone wire using a modem and PPP (Section 5.9.1), and the names of the various computers along the way would suggest that it was carried over ATM (Section 5.8.2), Fast Ethernet (Section 5.6.5.2), Gigabit Ethernet (Section 5.6.5.3), and almost certainly SONET across America (Section 5.8.1). Traffic to Australia often travels via America as there is a lot of spare capacity across the Atlantic. Links across the Pacific are often carried by satellite, but this connection used Cable and Wireless's optical fibre link. You can look at the route of an Internet connection with a program called 'traceroute' (UNIX) or 'tracert' (DOS).

2

The User Perspective

2.1   User Data

Users are at the top of the communication system. The layering strategy discussed in the previous chapter means that they only see the system through the application, and are insulated from the detailed workings. This has an interesting side effect – it allows the communication system to manipulate the data being transmitted for the user so as to put it in a more efficient form for transmission. This is termed source coding. Before looking at the detail of source coding, it is necessary to consider the underlying principles of information theory.

2.2   Introduction to Information Theory

2.2.1   Information Content

Information is the knowledge of something – whether the 'something' has to be true is a matter for the philosophers. In our case, we will simply deal in terms of messages, made up of one or more *symbols*. Something that produces these messages is called a *source*. The first attempt to mathematically quantify what is meant be information was undertaken by Hartley in 1928. His approach was as follows. Let a source produce a single symbol s which comes from an overall set of possible symbol values $\{s_1, \ldots, s_n\}$. Let the information given by that symbol be i_1. If the source produces l symbols, then intuitively, the resulting information, i_l should be $l \times i_1$. Hartley related information content to the number of possible outcomes, since he felt that the larger the number of possibilities, the more information was passed by knowing which one occurred. A single symbol can have n possible values, and if we have l of them, that results in n^l possible combinations. If we say information is a function of the number of outcomes, we have:

$$i_1 = f(n), \text{ and}$$
$$i_l = f(n^l).$$

However, we have $i_l = l \times i_1 = l \times f(n)$, so $f(n^l) = l \times f(n)$.

The only function which satisfies both these equations is the logarithm, so Hartley defined his measure of information of a source with n different symbols as $I = \log n$. Logs to the base of 2 are almost always used which give units of a *bit*, although a *nat* is the unit if natural logarithms (to base e) are used. This measure means that the information given by l such symbols is $\sum_l \log_2 n = l \log_2 n (= \log_2 n^l)$. Hartley's definition is sometimes written as $I = -\log_2 P(S)$, where $P(S)$ is the probability of the occurrence of one symbol, and all symbols are equally likely.

This definition is simple but only works when all messages are equally likely. This is usually not the case. We need to include a concept of probability because intuitively information depends on probability. If an event is either very likely, or not very likely, we can predict what the outcome usually is. This means that the information gained from actually knowing the outcome is less. To come up with a improved mathematical representation of information which takes this into account, we must change our view slightly so that instead of saying information is knowledge, we say information is the removal of uncertainty. We have a well-defined way of measuring uncertainty – probability theory. If we define a measure of uncertainty, we can use that to define information. This is the approach used by Shannon in 1948.

An uncertainty measure (let us call it $H(x)$) should have the following properties:

- If two experiments have equiprobable outcomes, the one with the greater number of possibilities has the greater uncertainty, i.e. $H(x_1) > H(x_2)$ if experiment x_1 has more equiprobable outcomes than x_2. For example, you are more uncertain about the number on the face of a die (6 possibilities, each with probability $\frac{1}{6}$) than you are about the face of a coin (2 possibilities, with probabilities $\frac{1}{2}$). Note that this corresponds in uncertainty to Hartley's information definition.
- If we have two independent experiments, the uncertainty about the answer to both experiments should be the sum of the uncertainty of each experiment individually.
- If the outcome of an experiment is grouped into two groups, the uncertainty of the experiment as a whole should be the sum of the uncertainties of each of the groups weighted by the probability that that group contains the outcome.

The first two conditions require $H(x)$ to be based on logarithms as before. The third condition required Shannon to add a probability factor. He defined the uncertainty, or *entropy* of an event X which can have n possible outcomes with probabilities $p_1, \ldots, p_n$, as $H(X) = -\sum_k p_k \log_2 p_k$. Entropy, which is the uncertainty of event X, has the following properties:

- It is a maximum when each possible outcome is equally likely.
- The ordering of the various probabilities does not matter.
- Entropy equals zero only when there is no uncertainty (i.e. when the probability of one of the events is 1).
- Adding an event with probability 0 to the set of possibilities does not alter the entropy.
- Entropy increases with the number of possible outcomes (the event of throwing a die has more entropy than tossing a coin).
- Entropy is a continuous function of its arguments.
- The entropy of two independent events is equal to the sum of the entropies of the individual events.
- If the set of possible outcomes is split into groups, then the values of the entropies for the individual groups, multiplied by their statistical weights, gives the original entropy.

Some of these properties are shown in Figure 2.1, which shows the entropy of a binary event (i.e. one with only two possible outcomes) for different probabilities. It has a maximum at 0.5 (i.e. it is as likely that the event will occur as it will not, and is zero both for the event always occurring and never occurring.

Entropy and information are related as follows $H(X) = -\sum_k p_k \log_2 p_k = \sum_k p_k I(X = x_k)$, i.e. entropy is the average information given by an event.

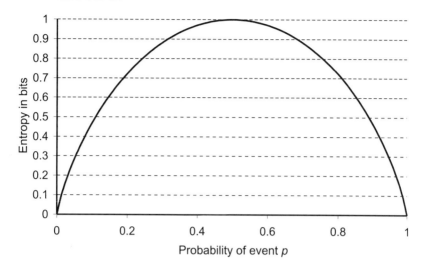

Figure 2.1 Entropy of a binary event for different probabilities

2.2.2 *Mutual Information*

Information is the removal of uncertainty. This means that strictly speaking when we talk about information we should consider two events – a 'before' and an 'after' or a 'source' and a 'destination' – and consider the information one gives about the other. Putting it another way, how much less uncertain we are about the first event with our knowledge of the second. The term *mutual information* is often used to emphasise that two events are required.

Consider two events X and Y. The information about X conveyed by Y to be $I(X;Y) = H(X) - H(X|Y)$. $H(X)$ is the uncertainty of event X. $H(X|Y)$ is the uncertainty of X given that Y is known, and how much lower this is compared to the original uncertainty defines the information. If X and Y are independent, $H(X|Y) = H(X)$, so $I(X;Y) = 0$, and Y gives no information about X. Note that $I(X;Y) = I(Y;X)$. This relationship is shown graphically in Figure 2.2.

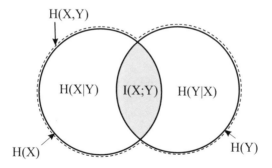

Figure 2.2 Relationship between entropy and information

2.3   Information Sources

A source is a producer of information symbols. A memoryless source is a particular type of source where the probability of a symbol being produced is independent of any previous symbols. If the probability of each symbol is unchanged over time the source is said to be stationary. Therefore, for such a source:

$$P(S_i = a_k) = p_k \forall i$$

A string of symbols produced by such a source is $S_1 S_2 S_3 \ldots S_n$ with all the symbols belonging to the same probability distribution. The entropy of this source is defined as:

$$H = -\Sigma_k p_k \log_2 p_k$$

(where the sum is taken over all k where $p_k > 0$). The set of valid symbols is called an *alphabet*.

A coin or a die forms a memoryless source since each toss is independent of the previous ones. If there is some dependency on previous outcomes, the source is said to have memory. Most real-world sources have memory. For example, characters on a page have memory because they are not independent. A 'q' will almost always be followed by a 'u', for example.

For sources with memory, it is usually not individual symbols, but rather groups of symbols, which form distinct messages. In the English language, each letter is a symbol, but only some groups of letters form valid combinations. The entropy of such a source is calculated over the messages rather than the symbols. If some of the possible combinations of symbols do not form valid messages, they are said to be *redundant* ('thear' for example).

Calculating the entropy of sources with memory requires considering groups of dependent symbols together to the point where the resulting messages can be considered independent, and then the entropy is calculated over the messages and not the symbols. For some sources this is extremely difficult, because dependency exists over large groups of symbols. Again, using the example of English, individual words or even sentences are not independent. In such cases the entropy of an individual symbol can be estimated by considering how easy it is to guess the following symbol based on the last. Over large pieces of English text, people can guess the next letter correctly slightly less than half the time, resulting in an entropy of just over one bit per letter.

2.4   Coding

Sources do not usually produce messages individually in a convenient form. A spoken language, for example, consists of several hundreds of thousands of words (in the case of English many millions), and having a separate symbol for each is not practical. A solution is to map messages on to a more manageable set of symbols. This is called encoding.

The code alphabet may be the same or different from the source alphabet. It is often, but not necessarily, binary.

2.4.1   Properties of Codes

A code is *distinct* if all the codewords are different. It is *uniquely decipherable* if any finite string from the encoded string corresponds to at most one message. A code is *instantaneous*

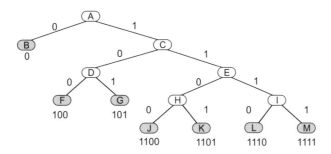

Figure 2.3   Code tree

or a *prefix-free code* if no codeword is a prefix of any other, which also makes it uniquely decipherable. A decision tree can be used to design codes.

Kraft's inequality is a necessary and sufficient condition for an instantaneous code. It states that the sum of the alphabet size D to the power of the individual word lengths l_i must be less than or equal to 1 for an instantaneous code, i.e. $\sum_{i=1}^{N} D^{-l_i} \leq 1$. McMillan's inequality gives the same condition for a uniquely decipherable code, so a uniquely decipherable code only exists if an instantaneous code also exists, so the condition is often termed the Kraft-McMillan inequality.

A code tree can be used to design a code. Starting at a root, the tree is extended with a branch for each possible code symbol. Each node in the tree corresponds to a set of code symbols and therefore to a possible codeword. The code is uniquely decipherable if each source symbol is represented by a different node. The code is instantaneous if each source symbol is represented by a terminating (i.e. leaf) node. In Figure 2.3, node A is the root node. An instantaneous code could be formed from B, F, G, J, K, L and M, or a subset of these. The figure shows the code 0, 100, 101, 1100, 1101, 1110 and 1111. If one of the intermediate nodes (say D, for example), the code would not be instantaneous unless any codewords further down the tree from that node (i.e. F and G) were removed to make D a terminating node. An instantaneous code can be extended to form a new instantaneous code by taking any terminating node and adding leaf nodes to it. This adds $D - 1$ codewords to the code (since D new leaf nodes can be added and the original node can no longer be used as a codeword since it is no longer a terminating node.

2.4.2   Minimum Code Lengths

If a (memoryless) source has entropy H, then any uniquely decipherable code for this source over an alphabet with D symbols must have a length of at least $H/\log D (= L_{min})$, and a code exists with length $\leq (1 + H/\log D)$. This is called the Noiseless Coding Theorem, because it defines the shortest possible code which fully describes the source (and therefore does not produce any distortion or noise). For the special case of binary codes, the length must be greater than or equal to the entropy in bits.

2.4.3   Redundancy and Efficiency

The *redundancy* of a code is the difference between the average length of the codewords and the minimum length that could have been used as a proportion of the average length, i.e.,

$\frac{L_{av}-L_{min}}{L_{av}}$. The average length of the codewords is simply $L_{av} = \sum_i p_i \, l_I$, where l_I is the length of codeword i and p_i is the probability of that codeword.

A closely related concept is the *efficiency*, η of a code. This is L_{min}/L_{av}, and is normally expressed as a percentage. Note that the redundancy is $1 - \eta$, and if a code is 100% efficient it has no redundancy. For the binary case only, $L_{min} =$ H, so it is possible to simplify these equations to, redundancy $= (L_{av}\text{-H})/L_{av}$, $\eta =$ H/L_{av}.

2.4.4   Types of Coding

There are four types of coding, although many authors talk of only three. Mathematicians usually forget line coding, while electronic engineers tend to forget cryptographic coding.

- **Source coding**: To reduce (compress) the information into as few symbols as possible to save on transmission or storage resources. Number of source combinations > number of encoded combinations
- **Channel coding**: Expanding the alphabet to allow errors to be recognised after transmission through a channel which may corrupt the data. Number of source combinations < number of encoded combinations
- **Cryptographic coding**: To transform the message so that others may not understand it. Number of source combinations = number of encoded combinations (in general)
- **Line coding**: To transform the message so that it may be transmitted more easily through the medium. Number of source combinations = number of encoded combinations

2.5   Source Coding

If we want to transmit information efficiently, we want to encode the most likely messages with short codewords to keep the average message length down. Where we know the source characteristics, Shannon-Fano and Huffman coding strategies are very useful. When we don't know the source characteristics, other techniques, such as run length encoding (used in facsimile) and Lempel-Ziv (pkzip, etc.) can be used.

Before we can perform any coding of the information, it must be in a suitable form to manipulate. In some cases, the source presents the information as discrete symbols which can be encoded directly. However, the information signal which has to be sent is often analogue (such as an audio signal or the output of a transducer), and the analogue signal must first be sampled to convert it into individual messages.

2.5.1   Quantisation

While it is possible to transmit a signal in an analogue form, any manipulation of data requires some form of digitisation. For example, measurements of distance are inherently analogue, but people digitise them in order to communicate. The width of a one penny coin is 20.20... millimetres, which is an indefinite number, and immediately the number is terminated to some finite number of digits the measurement is digitised. Thus, 20.2mm and 20mm are sampled values of the width of the coin.

2.5.2 Quantisation Accuracy

When the analogue signal is sampled, it is *quantised*, i.e., estimated by its nearest discrete symbol value. If we take measurements of the width of a coin to a single decimal place, our sample values are 20.0mm, 20.1mm, 20.2mm, and so on. The difference between the different possible quantisation values is the quantisation step. The difference between the actual value and the quantised value is the quantisation error, ϵ. The maximum value of this error is half the quantisation step (see Figure 2.4).

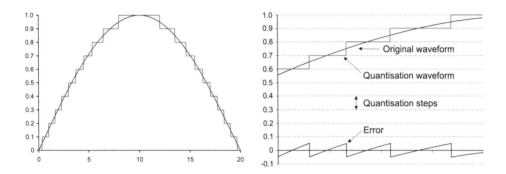

Figure 2.4   Quantisation noise

The Signal to Noise ratio (SNR) is the ratio of the power of the signal to the power of the noise. The noise in this case is the error due to quantisation. We do not know the exact value of an error on a particular sample, but we can make the reasonable assumption that all values in the input range are equally likely. For a quantisation step of d, the error therefore ranges over $(-d/2, d/2)$. Squaring and averaging this yields a noise power of $d^2/12$ (see Figure 2.5). The noise power is therefore proportional to the square of the quantisation step. Alternatively, if we have a signal with q quantisation levels, the quantisation error is proportional to the inverse of q^2. The higher the number of quantisation levels, the higher the SNR, but the more different symbols there are. The quantisation thresholds need not be uniformly distributed, and often are not. In sound systems, levels are spaced logarithmically, so there are more at low levels.

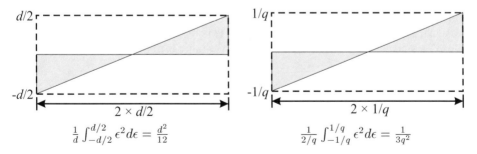

Figure 2.5   Quantisation noise calculation

2.5.3   Sampling Rate

For time-varying quantities like electrical signals, there is a further issue in addition to the accuracy of each sample – the rate of sampling. If the sampling rate is too low, we will not be able to reconstruct the signal from the samples.

Consider Figure 2.6. The original signal is the black line, and the samples shown will allow the signal to be reconstructed properly. However, if we take only every second sample, the result will be the grey line, which loses much of the detail of the original signal. The higher the sampling rate, the more samples are required, so more symbols must be transmitted. The key question is how few samples are required to completely describe the signal. Answering this question requires an understanding of the individual components that make up the signal. We can decompose a signal into different frequency components through Fourier analysis.

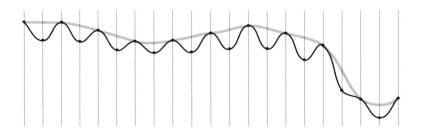

Figure 2.6   Waveform sampled at different rates

2.5.4   Fourier Series

This series allows the representation of almost any periodic waveform as a series of sinusoidal waveforms. The power of this approach is that if we then define the response of circuits to sinusoids of different frequencies, we can then calculate how these circuits will react to the waveform. The sinusoidal constituents of the waveform are called its *spectrum*, as is the *frequency-domain* representation of the waveform. This is a very powerful approach to the design of circuits. While the frequency domain representation does directly correspond to physical entities in the form of waves when we are considering electromagnetic radiation (radio waves, light, etc.), it is important to remember that the frequency domain representation of a signal is basically a mathematical model. This means that we can perform frequency analysis on signals which vary with other properties other than time, like space, for example.

2.5.5   Calculation of the Frequency Spectrum

Almost any period waveform can be expressed as a sum of a Fourier series. There are two main forms of the Fourier series – the exponential form and the trigonometric form. They are equivalent, given that $e^{j\theta} = \cos\theta + j\sin\theta$. For the exponential form, the signal is formed from $v(t) = \sum\limits_{n=-\infty}^{\infty} c_n e^{j2\pi n f_0 t}$, with $c_n = \frac{1}{T_0} \int_{T_0} v(t) e^{-j2\pi n f_0 t} dt$. For real signals, we can also use the trigonometric form, $\frac{1}{2}a_0 + (a_1 \cos x + b_1 \sin x) + (a_2 \cos 2x + b_2 \sin 2x) + \ldots$. The coefficients are $a_n = \frac{2}{T_0} \int_{T_0} f(x) \cos nx\, dx$ and $b_n = \frac{2}{T_0} \int_{T_0} f(x) \sin nx\, dx$. For the trigonometric form, the period is usually taken to be 2π, which then gives $a_n = \frac{1}{\pi}\int_0^{2\pi} f(x)\cos nx\,dx$ and $b_n = \frac{1}{\pi}\int_0^{2\pi} f(x)\sin nx\,dx$.

As an example, consider a triangular wave with $y = x/2$ for the interval $-\pi$ to π. The first 3 terms of a Fourier series of the waveform are shown in Figure 2.7, along with the sum of the first two terms and of the first three terms. Even with only three terms the underlying shape of the waveform can be seen. Figure 2.8 shows the approximation when 30 terms are included.

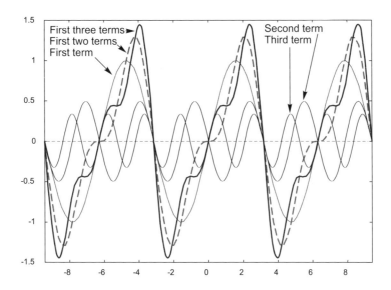

Figure 2.7   The first 3 terms of a Fourier series of a triangular wave with $y = x/2$ for the interval $-\pi$ to π, showing the individual sinusoids.

Although not every possible periodic waveform has a Fourier series representation, one condition which is sufficient (though not necessary) is that the integral of the square of the function be finite. This is the same condition as for a *power signal*, a signal with finite power over the period of the signal. All real signals must have finite power, so all real-world signals have Fourier series representations.

The Fourier series of a function is continuous. If the original signal was discontinuous, the Fourier series exhibits *Gibbs phenomenon*, where the Fourier series converges to the midpoint of the discontinuity. The Fourier series overshoots the ends of the discontinuity by about 18%, with oscillations which have the same frequency as the highest component in the Fourier series. This overshoot, called 'Gibbs ears', reduces in energy as $N \to \infty$, but the amplitude does not reduce. The triangular wave of Figure 2.8 has a discontinuity, and so Gibbs ears are present, but they are easier to see on a rectangular wave such as Figure 2.9). Note that no real signal has a discontinuity – this would require that electrons move with infinite speed, which is not possible. However, the mathematical waveform we use to represent the real wave may have a discontinuity in order to approximate the signal. An example is a step function.

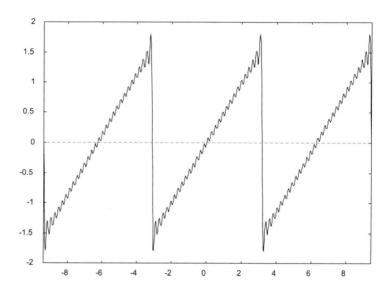

Figure 2.8   The first 30 terms of a Fourier series of the triangular wave shown in Figure 2.7

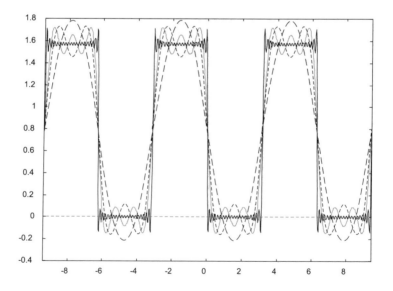

Figure 2.9   Fourier series of the function $y = 0, 0 \leq x < \pi, y = \pi/2, \pi \leq x < 2\pi$. The series with
2, 3 and 4 terms are shown, along with the series for all terms up to $n = 40$.

2.5.6   Frequency Spectrum

For periodic waveforms, we can use the terms in the Fourier series to define a frequency spectrum for the waveform. This is a graph of the frequency components and their amplitudes. The frequency spectrum of a periodic waveform is called a line spectrum, since it has components at $\frac{1}{T}$, $\frac{2}{T}$, $\frac{3}{T}$, etc (where T is the period of the waveform). The lowest of these frequencies, $\frac{1}{T}$, is called the *fundamental frequency*.

A line spectrum only results from signals which are truly periodic, i.e. they extend from $-\infty$ to ∞. While real signals do not do this, we can often consider them truly periodic within the time domain we are considering.

The question of how to treat signals which are not periodic now arises. In this case, we can calculate their frequency response using the Fourier Transform. The Fourier transform is defined as:

$$F(f) = \int_{-\infty}^{\infty} v(t)e^{-j2\pi ft}dt$$

Notice that the Fourier transform is a continuous function, and therefore an aperiodic waveform has a continuous spectrum, rather than the line spectra seen with a periodic waveform.

Two specific transforms are worthy of note. The first is the transform of an impulse, i.e. the delta function, $\delta(t)$, which equals 1 at $t = 0$, and equals 0 elsewhere. Its Fourier transform is

$$
\begin{aligned}
F(f) &= \int_{-\infty}^{\infty} v(t)e^{-j2\pi ft}dt = \int_{-\infty}^{\infty} \delta(t)e^{-j2\pi ft}dt \\
&= \int_{t=0} 1e^{-j2\pi ft}dt + \int_{t\neq 0} 0e^{-j2\pi ft}dt = \int_{t=0} 1dt = 1
\end{aligned}
$$

More generally, $F(A\delta(t)) = A$, so the frequency spectra of an impulse is a constant containing all frequencies.

The second function worth remembering is the spectrum of a square pulse. Consider a pulse with amplitude A from $-T/2$ to $T/2$, with value 0 elsewhere.

$$
\begin{aligned}
F(f) &= \int_{-\infty}^{\infty} v(t)e^{-j2\pi ft}dt = \int_{-T/2}^{T/2} Ae^{-j2\pi ft}dt = \frac{-A}{j2\pi f}\left[e^{-j2\pi ft}\right]_{-T/2}^{T/2} \\
&= \frac{2A}{2\pi f}\sin\left(\frac{2\pi fT}{2}\right) = AT\,\mathrm{sinc}\,(fT)
\end{aligned}
$$

The relationship between the frequency domain and the time domain is emphasised by the fact that a constant of amplitude A has a transform of a delta function of value A at $f = 0$, i.e. a dc component. Also, a time domain sinc pulse has a (scaled) rectangular response in the frequency domain.

Some properties of Fourier transforms are shown in Table 2.1. Multiplication in the time domain equates to convolution in the frequency domain, and vice versa. Repetition in the time domain equates to sampling in the frequency domain (with a scaling factor), and vice versa.

2.5.7   Minimum Sampling Rate

From the above, sampling a signal in the time domain results in the frequency domain representation repeating at the sampling frequency. Therefore, if the sampling frequency is more than twice the bandwidth of the signal, we can reconstruct the original signal, because the repetitions will not overlap with the original signal. Therefore, if the analogue signal has a bandwidth of W, we require to sample at a rate of at least $2W$ (or more).

Table 2.1 Properties of Fourier transforms

	$g(t)$	$F(g(t)) = G(f)$
Time scaling	$g(t/\tau)$	$|\tau|.G(f)$
Time shift	$g(t - \tau)$	$G(f)e^{-j2\pi f\tau}$
Duality	$G(t)$	$g(-f)$
Multiplication	$g(t)h(t)$	$G(f) * H(f)$
Convolution	$g(t) * h(t)$	$G(f)H(f)$
Repetition	$\sum_{n \in Z} (g(t) * \delta(t - n\tau))$	$\left|\frac{1}{\tau}\right| \sum_{n \in Z} \left(G(f) \times \delta\left(f - \frac{n}{\tau}\right)\right)$
Sampling	$\sum_{n \in Z} (g(t) \times \delta(t - n\tau))$	$\left|\frac{1}{\tau}\right| \sum_{n \in Z} \left(G(f) * \delta\left(f - \frac{n}{\tau}\right)\right)$

Reconstruction of the signal is undertaken by removing the high frequency copies of the frequency spectrum with an ideal low pass filter. A rectangular frequency response corresponds to a *sinc* function in the time domain. Note that since this function has an infinite response, all physical systems will have some distortion.

2.5.8 Pulse Code Modulation (PCM)

Pulse code modulation is the name given to sampling an analogue signal and replacing it with discrete samples of the signal at instants of time. The more samples that are taken, the higher the symbol rate. Also the more accurate the samples, the larger the number of possible symbols. The symbols are often strings of binary digits, so that larger number of samples and the greater sample accuracy translate to higher bit rates.

2.5.9 Source Coding for Memoryless Sources

2.5.9.1 Shannon-Fano Encoding

List all the messages in descending order of probability, then divide the table into sections with as equal probability as possible (two sections for a binary code, three for tertiary, etc). Assign a 0 to the first part, 1 to the second, etc. Continue until there are no sections left to be divided.

2.5.9.2 Huffman Encoding

Huffman coding proceeds as follows. The message symbols to be coded are listed in order of probability, with the highest probability at the top. If the code alphabet is binary, the two branches with the lowest probabilities are then combined, and before the combination one is labelled 0 and one is labelled 1. The branches are then re-ordered by probability and the process repeats. This continues until there is only one branch. The codewords are then read off by reading backwards from the last point and adding any 0s and 1s which are encountered.

Take a system with the following probabilities: A 0.25, B 0.3, C 0.35, D 0.1. These are ranked a follows

C   0.35
B   0.3
A   0.25
D   0.1

The smallest are 0.25 and 0.1. These are combined into a single branch AD, and the branch to A labelled 0 and the branch to D labelled 1. The new ordering is

C   0.35
AD   0.35
B   0.3

We have a free choice as to whether to order AD above or below C. We get different but equivalent codes in each case.

AD and B are combined, AD being labelled 0 and B being labelled 1. This leaves

ADB   0.65
C   0.35

which are combined, with ADB labelled 0 and C labelled 1, to give only one branch, so we are finished. Extracting the codewords gives us:

A: 0 to branch ADB, 0 to branch AD and 0 to branch A, so the codeword is 000
B: 0 to branch ADB, 1 to branch B, so the codeword is 01
C: 1 to branch C, so the codeword is 1
D: 0 to branch ADB, 0 to branch AD, and 1 to branch D, so the codeword is 001

The coding process is shown in Figure 2.10.

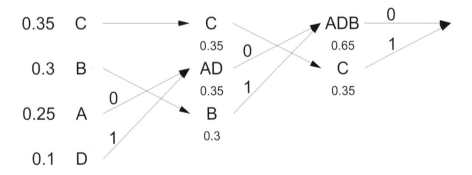

Figure 2.10 Huffman code construction example

In order to use Huffman coding for code alphabets with more than two symbols, we need to make a small modification to the grouping stage. Instead of combining the smallest 2

Table 2.2   Main types of compression

	Repetitive sequence suppression	Zero suppression
Symbol based coding (Computer Sc. Entropy coding) *Lossless*		Run length encoding
	Statistical encoding	Huffman/Shannon-Fano
		Pattern substitution
		Arithmetic
Syntax based coding (Computer Sc. Source coding) *Lossy (though some differential schemes* can be lossless when combined with a symbol based scheme)*	Transform encoding	FFT
		DCT
	Differential encoding	Differential PCM *
		Delta modulation
		Adaptive Differential PCM *
	Vector Quantisation	CELP (Voice)
		Fractal (Image)

probabilities before moving to the next stage, we combine the smallest n, where n is the number of symbols. Then when we number the different branches of the tree we use a different symbol for each of the branches which divide at that point. This means that we remove $(n-1)$ inputs at each stage. Since we are left with 1, before that we must have $1 + (n-1)$, before that $1 + 2(n-1)$, and so on. If the number of inputs we start with is not equal to 1 (mod $n-1$), we add additional dummy inputs with probability 0. The number of additional inputs is the smallest number which will mean that the new total is congruent to 1 (mod $n-1$).

2.5.10   Source Coding for Sources with Memory

Compression using Huffman and Shannon-Fano algorithms works well for sources producing symbols for a fairly restricted alphabet, but they do not work well when the alphabet is very large and each symbol is used only a few times. For example, a picture may consist of many millions of pixels, each of which may have a slightly different colour. In this case we may be able to make use of the memory of the source. Other types of compression have therefore been designed to address these situations. Table 2.2 shows the main types of compression.

Compression is lossless if the source can be recreated exactly from the compressed data. Lossy compression throws away some information in the compression process, so the source cannot be exactly reconstructed. An example is reducing the number of distinct colours in an image. The compressed version of the image will be different, but assuming the compression system is properly designed, it should be perceived by the user to be the same.

2.5.10.1   Zero Suppression

This is where one commonly occurring symbol is replaced by a token to represent the symbol and the number of times it is repeated. Other symbols are passed through unchanged. This symbol chosen is often '0' – hence the name – but need not be.

2.5.10.2   Run Length Encoding

This is an extension of this where runs of any symbol can be replaced by the symbol and a repeat symbol specifying the number of occurrences. A pure black and white image, for example, can be encoded by simply sending the number of white pixels, followed by the number of black, followed by the number of white, and so on. 11111110000000110000010000000000000010 becomes 07625119021. The block of eleven is split into 902 as 9 is the largest symbol value. '11' would be confused for 1 followed by 1.

2.5.10.3   Pattern Substitution

This is where common sequences of symbols are replaced with a *meta-symbol* to represent the whole sequence. Acronyms are a form of pattern substitution, for example, EU for European Union. Two widely used algorithms for pattern substitution were developed by Ziv and Lempel in 1977 and 1978. In their 1977 algorithm, the first few symbols are transmitted without compression, and then the following symbols are checked against the previously transmitted set. If any pattern has been sent, the position and length of the sequence are sent rather than the sequence of symbols itself. The 1978 algorithm improved on the 1977 one by starting with a dictionary containing all the symbols from the source alphabet, and then adding to it as strings are transmitted.

2.5.10.4   Differential Compression

This encodes the difference between the previous symbol and the current symbol rather than the symbol itself. Where there is a relationship between symbols (i.e. samples of a waveform), the difference may be significantly smaller than the absolute value, and so easier to encode. This can be improved by using adaptive differential encoding, where the coder predicts the next symbol, and sends the difference between the prediction and the actual value. A simple form of differential encoder is a delta modulator, which only sends a single bit to increase or decrease the output.

2.5.10.5   Transform-based Compression

The aim of transform coding is to convert the signal into another domain where it may be represented more efficiently, and compress it in this second domain. For example, a time domain signal could be transformed into the frequency domain, and if no high frequency components are present, fewer coefficients would be needed to send the information. A transform and its inverse will not cause any distortion to the signal unless the data is truncated in some way, so in this pure form transform coding is lossless.

   In general there will be some components throughout the range, even if they are small. Therefore, to introduce compression, insignificant components are discarded in a process called 'quantisation'. The larger the number of components which are discarded, the higher the compression, but the greater the distortion introduced.

2.5.10.6   Vector Quantisation

Vector quantisation replaces the source with a reconstructed signal consisting of components defined by the encoder. The CELP (codebook excited linear predictive) speech coder used in mobile radio has a very low bit rate (down to 2.4kbit/s) and has a codebook of speech sounds which can be adjusted by changing parameters. These parameters are sent, rather than the sound itself. Fractal compression of images uses the same technique where fractals – arbitrary shapes which appear in various forms in the image – are sent with instructions on how to use them to create a similar image, rather than an exact copy of the image itself.

2.5.10.7   Compression Examples

- **Fax**: Group 3 fax uses run length encoding and Huffman encoding. The scanned image contains 1145 lines of 1728 white and black pixels. Each line is run length encoded and then Huffman encoding is used on the resulting numbers. The original two million bits are replaced by about a quarter of a million bits by this process.
- **Text**: Lempel-Ziv algorithms work well on text, usually achieving a reduction of 20 to 25%. Note that this is not a reduction to near the true entropy of the speech, because the algorithm only considers the symbol sequence and not things like spelling and grammar.
- **Speech**: The standard ITU speech coder has a bandwidth of 4kHz, with 8000 8 bit samples/sec (64kbit/s). Modern CELP coders can achieve the same sound quality (called toll quality) with a 8kbit/s rate. However, because CELP uses a code book of sounds matched to speech, the quality of other sounds is poor. The ITU code would work on any sound source up to 4kHz.
- **Images**: Still images often use GIF, which reduces the colour depth to 256 colours, and then uses a Lempel-Ziv algorithm to achieve lossless compression. JPEG uses a discrete cosine transform on blocks of 8 by 8 pixels. It reduces the resulting components by truncating the unimportant (high frequency) components ('quantising'), using differential encoding on the most important (DC) component, using run length encoding, finally Huffman or arithmetic encoding to compress the remaining components as much as possible. MPEG uses a Finite Fourier Transform along with a system of reference and predictive frames to reduce bit rate as much as possible. For the predictive frames, only the differences are transmitted. MPEG-1 gives VCR-like quality in just over 1Mb/s. MPEG-2 gives broadcast quality in about 2Mb/s, and is being used by the new digital TV services. MPEG-4 is designed for very low bit rates (< 64k bit/s) for video conferencing. Very, very low bit rates are possible.

2.5.10.8   Practical Compression – JPEG

A good example of a practical compression is the JPEG image compression standard, which uses transform, differential and run length encoding in order to compress images as much as possible. Images are split into blocks of 8 by 8 pixels, and then a 2 dimensional Discrete Cosine Transform is used on each of the blocks. If the image is a colour image, the transform is applied to each colour channel separately.

In JPEG, quantisation is performed by dividing each of the coefficients in the 8 by 8 array by a predefined value which favours the low frequency components, and leaving only the integer part of the division. This means that the bottom right-hand part of the array will fill with zeros. The most significant component will be the top left component, which represents the 'dc' value, the average value in the 8 by 8 block. To introduce greater compression, differential encoding is used of this value, and then the coefficients are read out in a diagonal fashion so that zero values are grouped together as much as possible. Finally, run length encoding is used to compress the runs of zeros.

2.6 Questions on the User Perspective

2.6.1 Questions on Information Theory

1. What is the information content (in bits) of the following events:
   (a) Tossing a coin?
   (b) Tossing two dice?
   (c) Drawing a card from a shuffled deck of cards: considering each card individually; considering each suite separately, considering each card number independently of suite, and considering only the colour?

2. By making reasonable judgements regarding image resolution, estimate the information content of a PAL TV picture.

3. Calculate the entropy of the event of tossing a pair of dice:
   (a) considering the individual scores of each die (so (1,6) and (6,1) are separate events);
   (b) considering the scores without regard to number (so (1,6) and (6,1) are considered to be the same);
   (c) considering only the sum of the scores of the dice.
   (d) On average, how much information does the sum give about the individual scores of the dice?

4. In the television game show, *Who Wants to be a Millionaire?*, contestants have to pick a single correct answer from four choices A, B, C and D. At one point during the game, they are allowed to go '50:50', where two incorrect answers are removed, leaving the correct answer and one wrong answer from which to choose.

   (a) Assuming a contestant has no idea what the correct answer is, and so considers each outcome equally likely, how much information is given by going '50:50'?
   (b) Alice is playing the game and has become stuck on a question. She is fairly certain the answer is A or B, and hopes that by playing '50:50' she will remove one of these. She estimates that it is 45% likely that the answer is A, 40% likely the answer is B, 10% likely the answer is C and 5% likely the answer is D. Calculate the entropy based on her assumptions before and after '50:50' is played if going '50:50' removes B and D.
   (c) Calculate the entropy based on her assumptions before and after '50:50' is played if going '50:50' removes answers C and D.
   (d) What is the information (in bits) given by going '50:50' in each of the above two cases?

5. By expanding $H(A, B)$ in terms of probabilities and using Bayes Theorem, show that $H(A, B) = H(A|B) + H(B)$.

2.6.2 Questions on Memoryless Source Coding

1. A memoryless information source produces 8 different symbols with respective probabilities of 1/2, 1/4, 1/8, 1/16, 1/32, 1/64, 1/128, 1/128. These symbols are encoded as 000, 001, 010, 011, 100, 101, 110, 111 respectively.

(a) What is the entropy per source symbol?

(b) What is the efficiency of this code?

(c) Design a code using the Shannon-Fano algorithm, and calculate its efficiency.

(d) Design a code using the Huffman algorithm, and calculate its efficiency.

(e) If the source symbol rate is 1000/sec, on average what is the encoded bit rate?

2. A memoryless information source generates symbols with probability 0.65, 0.2, and 0.15.

(a) Calculate the entropy per symbol.

(b) Calculate the probabilities of all possible messages consisting of two symbols, and the corresponding entropy. Compare this to the previous result.

(c) Calculate the redundancy of the information source.

3. Design a code to encode a memoryless source with six symbols using a tertiary encoded alphabet $\{0,1,2\}$. A sample of the source output is ACAAABEBCDAEABDCAEFAABDF. What is the efficiency of your code? Compare it to the code $\{$A:00, B:01, C:02, D:10, E:11, F:12$\}$

4. Remote sensors can have two states – 'SET' and 'CLEAR'. On average, 99% of sensors are in the 'CLEAR' state.

(a) A one bit code, $0 \rightarrow$ CLEAR, $1 \rightarrow$ SET, is used. What is its efficiency?

(b) An improvement of the efficiency is sought by constantly taking two messages together. Determine a suitable binary code for this, and calculate its efficiency.

(c) Repeat for groups of 3 symbols.

(d) Repeat for groups of 4 symbols.

(e) Comment on any trends you see in the above results.

2.6.3   Questions on Source Coding for Sources with Memory

1. Encode the following sequence using delta modulation : 1, 2, 4, 4, 4, 4, 5, 6, 6, 5, 5, 4, 3, 1, 0. Assume the decoder is outputting 0 before reception of the first bit. What is the mean squared error?

2. A source generates the sequence 15, 14, 12, 9, 8, 8, 10, 11, 12, 14, 15, 15, 14, 12, 10, 9.

(a) Encode the sequence using differential PCM

(b) If you are told that the values which will be received are integers in the range 0 and 15, how many bits per symbol will be required in differential PCM? Justify your answer.

(c) Is the system you propose lossy or lossless? If lossy, how could you make it lossless?

3. The same sequence as given above is to be transmitted using adaptive PCM. The estimate used of the next sample is to assume that the difference between the current value and the next is the same as between the current and the last.

(a) What are the values to be sent?

(b) Estimate the number of bits required per symbol.

4. Use the 1977 Lempel-Ziv algorithm to compress the sequence ABCCDABDCCBC-CBECDA. Compare the number of bits before and after compression. The sequence is 18 symbols long, so 5 bits are needed to address the string. Comment on what happens regarding addressing long sequences. If the algorithm could only address previous strings over a finite history using a sliding window, what advantage and disadvantage would this have?

3

The Security Perspective

3.1 Introduction

Cryptographic coding has a number of aims. The most obvious one is to ensure that the encoded information is kept secret from all but the intended recipients. However, since the communicating parties may be physically separate, there are other security requirements. These include ensuring that the other party is who they claim to be and can be held to any agreements made. A second requirement of cryptographic coding is therefore to ensure communicating parties do not cheat

- by not playing by the rules
- by pretending to be someone else
- by obtaining information they are not supposed to have
- by falsely denying a transaction which has taken place.

3.2 Types of Cryptography

The relationship between the various categories of cryptography is shown in Figure 3.1.

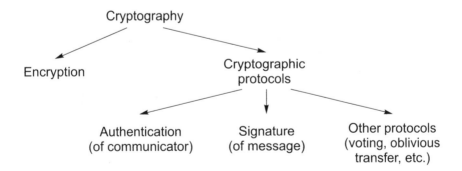

Figure 3.1 Categories of cryptography

Encryption – scrambling or disguising the information – is used to protect the data from being read by unauthorised parties. There are two types of encryption: diffusion, where the information is hidden, and confusion, where the information is scrambled so that even although a message may be detected, its meaning cannot be ascertained.

3.2.1   Diffusion

Diffusion is also known as 'steganography', the art of secret writing. Classical techniques included invisible ink, or putting tattoos on the scalp of couriers, waiting for their hair to grow, and then sending them to the recipient.

Diffusion is still used today. For example, specific letters or letter combinations may be used in an email which only have meaning for the intended recipient. Other readers would simply see an innocent email. This sentence also appears innocent, being text advocating a key idea. However, the second letter of each word of the last sentence passes a different message – 'help needed'. Spread spectrum radio communications, which is used by the military, is a form of diffusion. The frequency the radio signal is transmitted at is changed rapidly in an unpredictable fashion which is known only to the intended receiver. The technique is called 'frequency hopping', and if the receiver moves (or hops) to the correct frequencies it will receive the entire message. An eavesdropper listing in on one frequency would receive only a tiny part of the message, and may therefore not even know that a message was being transmitted.

Steganography can also be used to send information within sound or video files. Occasional bits in the picture could be altered by the addition of the information to be transmitted. The resulting picture, while containing the hidden information, would look (or sound, in the case of an audio file) the same to a viewer or listener. In Figure 3.2, the line drawing on the left is hidden within the photograph on the right. The photograph has 16 bit grey scale values, the least significant bit of which is used to hide the line drawing.

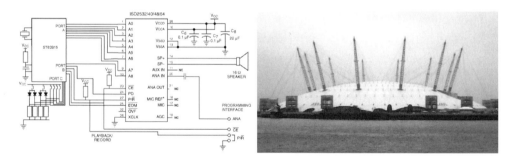

Figure 3.2   Line drawing hidden within a photograph

The same technique can be used for copy protection. Images could be watermarked with a pattern which is invisible to the viewer but which can be detected as a pattern in certain pixels. This could be used to identify if a picture was copied from an original source. Sophisticated watermarking schemes allow the watermark to remain even if the picture is altered, and the system also ensures that if a picture is modified, the modification can be detected. This is important for pictures from police speed cameras, for example. An older example of this technique is the practice of adding spurious words to dictionaries so if the dictionary is copied, the spurious word will be detected and the fact that the dictionary was copied is ascertained. Chip designers often add small pieces of redundant logic to their designs for a similar reason in case their mask design is copied.

A significant problem with diffusion is that once the method is known, it is usually trivial to attack. If diffusion is used to transfer information between a number of people, the secret is bound to come out. While a number of programs are available to hide information within images, for example, these programs can also be run speculatively on images to see if any information is hidden. In order to safeguard the information, it is necessary to further encrypt this hidden information, but that involves the use of the second method, confusion, discussed below.

Another problem with diffusion is that a large amount of data is required to obscure a small message. In order to hide a tree in a forest you need a forest! If one bit in every frame is used to transmit hidden information, the number of remaining bits must be sufficient so that the covert bits cannot be detected. This would imply a raw file size at the very least about 8 times as large as the data being hidden, and perhaps significantly more.

3.2.2   Confusion

The second encipherment method is 'confusion', where the message is altered so that even though its presence is detected, it cannot be deciphered. There are two basic methods of confusion: transposition, where the position of individual message symbols is changed, and transformation, where the symbol values are changed. Both methods can be used at the same time.

We have already come across the concept of a code – a set of rules which transform each input message into a defined set of symbols. This may be done to make transmission easier (line codes) or to detect or correct errors (error correcting codes). A *cipher* is a code whose purpose is to disguise the message being sent. Ciphers can be sub-divided into *block* ciphers and *stream* ciphers. In a block cipher, each part, or block, of the message is encrypted independently of the others. In a stream cipher there is some form of feedback so that the encryption of message symbols depends on the symbols previously encrypted.

The encryption algorithm consists of two parts: a secret part and an open part. The success of the system depends on the secret part remaining secret, so it is usual to try to limit the amount of secret information required. The secret information is called the *key*. The algorithm itself should be within the open part since it will be known to everyone using the system (as opposed to the key which will be unique to the parties actually communicating). Given the number of people having access to the algorithm, it must be assumed to be known, and in fact there are some advantages in this approach since flaws in security are more likely to be discovered in published algorithms than in those which are kept secret, and which are therefore only studied by those with malevolent intent.

3.3   Cryptographic Scenarios

Communication occurs between a source and a destination. In the cryptographic community, rather than refer to a source A and a destination B, it is the convention to use a scenario where the source, named Alice, is trying to communicate with a destination, Bob. The channel between them is open to attack.

There are two options for the attacker (variously called 'Charles' or 'Eve' depending on the author). The first attack is the *wiretap channel*, where Eve can listen to messages passing between Alice and Bob but cannot alter them, and the second is the substitution attack, where Eve can also alter messages or substitute them (see Figure 3.3). The first form of attack is relatively simple for most transmission media, especially for radio where no physical connection is required, whereas the increasing use of packet switching networks, where data is processed at each router, has meant that the interception attack is often possible as well. An Internet Firewall is a good example of the second type of channel in action.

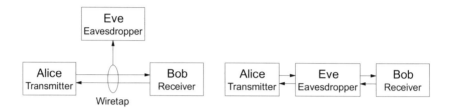

Figure 3.3   Wiretap channel and intercept channel

As well as attacks on cryptographic protocols which we will cover later, there are a number of types of attack on the encryption algorithm. These are:

- **Ciphertext only**: This is a simple wiretap, and Eve can only read encrypted messages as they pass by.
- **Known plaintext**: In this attack, Eve has the unencrypted text and encrypted text of some messages. Often at least part of the plaintext can be guessed, either being the address or name of the transmitter, or the time of transmission. This approach was used in the Second World War where bombs were dropped on a port and encrypted messages reporting this recorded in the hope that the name of the port and other details of the attack would be included.
- **Chosen plaintext**: In this attack, Eve can submit messages to be encrypted, and read to encrypted text, either by fooling Alice into sending the messages for her, or by stealing the encryption equipment and using it.
- **Chosen ciphertext**: The final option is for Eve to be able to get Bob to decrypt the encrypted messages she generates, and that she can read the decryption. Again this could be done by stealing or replicating the encryption equipment.

In all attacks, it must be assumed that the algorithm itself is known. Quite simply, too many people know about any algorithm for it to be kept secret. This assumption is known as Kerchoff's Assumption. Security depends only on the key. Some commercial encryption software developers try to keep their algorithms secret for 'enhanced security', but reverse engineering computer code is relatively easy, and it is unreasonable to assume such codes will not fall into the wrong hands.

3.4   Private Key Systems

3.4.1   Introduction

Private key systems are also called symmetric systems because both communicating parties have the same key, and decryption is simply the reversal of encryption through the use of the key. The most significant advantage of private key systems is that they have relatively low complexity for a given level of security.

3.4.2   Transposition Ciphers

In transposition ciphers, the positions of message symbols are altered, but the values are unchanged. A simple cipher is a 'rail fence', where a message is read in to a grid one way and then read out another. Figure 3.4 shows this in operation. The mesage is read in to the grid in columns and read out in rows. An ancient variant was to write a message on a strip wound round a cylinder, which was then unwound and sent by messenger. Both forms are simply interleaving, as covered in Section 5.4.5, only used for a different purpose. All that an eavesdropper needs to do is to find the depth of the grid (or diameter of the cylinder). Increasing the depth of the grid causes the same problems of delay as it does for interleaving for error control.

T	H	I	S	I	S	A
M	E	S	S	A	G	E
T	O	S	E	N	D	.

THI SI SAMES SAGE TOS END.

⇓

TMTHEOI SSS SEI ANSGDAE.

Figure 3.4   Simple transposition cipher

An improvement is to read the columns of the grid in a more complex order than simply left to right. A key word could be used to specify the ordering of the read out of the columns by the alphabetical ordering of its letters. For example, if the codeword was 'cipher', we would read into 6 columns, and then read out column 1 (c), 5 (e), 4 (h), 2 (i), 3 (p) and 6 (r). However, the system is still very susceptible to trial and error attacks.

3.4.3   Transformation Ciphers

3.4.3.1   Mono-alphabetic Cipher

The simplest type of transformation cipher is to take each symbol in the message and change it to a different symbol in the ciphertext. The symbols in the ciphertext will usually come from the same alphabet as the message, but this need not be the case. Since a given message symbol is always transformed to the same ciphertext symbol, the system is called mono-alphabetic.

An early use of such a system (though by no means the first recorded use of cryptography, which dates back to ancient Egypt and was used in recognisably modern forms by the Greeks) was reported by Suetonius, an ancient Roman gossip columnist, who wrote that Julius Caesar used a system of transformation when corresponding with his friends. He replaced each letter with one three further on in the alphabet (see Figure 3.5), proving not only the need for secure political communications, but the difficulty of keeping such communications secret from the press! The term 'Caesar Substitution' is now applied to any cipher with such a shift between the message and cipher alphabet, even if the shift is not three.

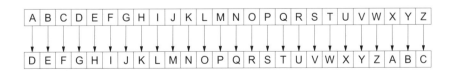

Figure 3.5   Caesar substitution cipher

More secure, but only slightly more so, is an arbitrary substitution, where the order of the letters in the substitution is changed. However, although such a system has more possible keys (26! rather than 26 possible in the Caesar system, one of which is trivial), the problem with all monoalphabetic transformation ciphers is that they are very easy to attack using frequency analysis. The redundancy inherent in the English language is such that only about 25 letters of ciphertext are required in order to decrypt the message. If spaces remain in the ciphertext, decryption is even easier, since a single letter word can only be 'a' or 'I'. Other information from the message may leak through. A shareware 'security' program used Java to encrypt web site addresses unless the correct key was entered. However, the ciphertext was easy to intercept, and since all web URLs start 'http', and most contain 'www' and end in '.html' or '.htm', the encryption of h, t, p, w and m were usually given. The program designers made life easier for the hacker by leaving the punctuation unencoded, so the '.xyz' at the end of the host name was almost certainly '.com'. This left so few letters that the remaining possibilities could be checked by trial and error.

Another problem with an arbitrary substitution is the length of the key, since the transformation of each letter has to be specified. This is not easy to remember. A simple system, sometimes referred to as a Vigenère substitution (although in fact it has no connection with the 16th century French diplomat), is to use a code word to specify the first few substitutions, and then fill in the remaining letters. The key, in the form of the code word, is easy to remember, but the cipher is very poor, with letters late in the alphabet (such as w or y) unlikely to be enciphered at all (see Figure 3.6).

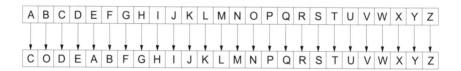

Figure 3.6   Vigenère substitution cipher

3.4.3.2   Polyalphabetic Cipher

One way to overcome the frequency analysis attack is to use different alphabets for the transformation depending on the position of the message symbol. The system, invented by Girolamo Cardano, is named after Vigenère, although he actually developed the autokey system described in Section 3.4.7. Blaise de Vigenère has the unfortunate distinction of being incorrectly associated with two ciphers he did not develop and not with the much superior cipher he did in fact invent. A codeword is used to define a number of Caesar substitutions which are then used for encryption. For example, let the codeword be 'code'. This defined four alphabets, one with a shift of 2, one with a shift of 14, one with 3 and one with 4. The encryption table and an example of encoding are shown in Figure 3.7.

Such polyalphabetic ciphers are better than monoalphabetic ciphers, but they are still vulnerable to attack using frequency analysis once the attacker calculates the repetition length of the codeword, since they can then perform frequency analysis on each alphabet individually. In the Vigenère system, since each alphabet is a Caesar substitution, attack is even easier.

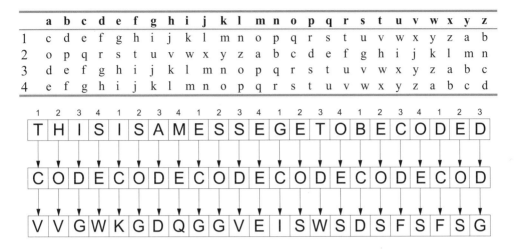

	a	b	c	d	e	f	g	h	i	j	k	l	m	n	o	p	q	r	s	t	u	v	w	x	y	z
1	c	d	e	f	g	h	i	j	k	l	m	n	o	p	q	r	s	t	u	v	w	x	y	z	a	b
2	o	p	q	r	s	t	u	v	w	x	y	z	a	b	c	d	e	f	g	h	i	j	k	l	m	n
3	d	e	f	g	h	i	j	k	l	m	n	o	p	q	r	s	t	u	v	w	x	y	z	a	b	c
4	e	f	g	h	i	j	k	l	m	n	o	p	q	r	s	t	u	v	w	x	y	z	a	b	c	d

Figure 3.7 Vigenère polyalphabetic cipher

3.4.4 One Time Pad

We can extend the notion of a polyalphabetic cipher to form the One Time Pad, which is the only completely secure cryptosystem. In the polyalphabetic cipher, the problem is the repetition of the key. In the One Time Pad, each message is encrypted with a key which is then discarded and never used again. Therefore, the key is only used once, giving the cipher its name. The cryptogram depends on the message and the key, but since the key is unique to that transmission and is never reused, the eavesdropper has no way of knowing what it is and breaking the cipher. Encoding can be very simple – simply adding the key to the message (see Figure 3.8).

Figure 3.8 One time pad encryption system

While this is the method of choice for any self-respecting spy, the difficulty is that the key is used up as fast as the message, and cannot be generated in any predicable way (or it may be possible to guess what the key is). The problem of transporting an encrypted message safely is replaced with one of transporting a key of equivalent length, and the problem of the generation of such a key. The system is only suitable for sending relatively small quantities of very high security data where the key can be transported securely off-line.

3.4.5   Shift Register Encoders

In Section 6.4.3, we consider the use of scrambling in line coding applications to try to break up long runs of zeros and ones. It is possible to use similar techniques for cryptography, by taking a Linear Feedback Shift Register (LFSR), initialising it to some value, and then adding its output to the message. A similar LFSR in the receiver will return the bits to their original sequence. An m stage shift register is, in general, capable of producing non-repeating sequences of length up to $2^m - 1$, which suggests good security even for relatively small values of m, but unfortunately the system is very susceptible to known plaintext attack. Since there are only m stages, only $2m$ bits of the shift register sequence are required in order to calculate the feedback taps and break the code.

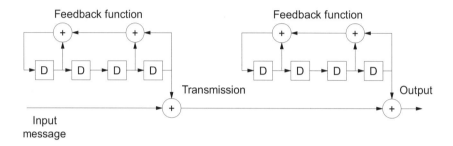

Figure 3.9   LFSR encryption system

Of practical application are non-linear feedback shift registers, which are robust against this attack. The non-linear element could be a JK flip-flop or a multiplexer. Multiplication can also be used, but the problem is that it is an asymmetric operation, producing 0 in three cases and 1 in only one (for 1×1), which means that there may be too many 0s in the output sequence and message characteristics may leak through.

The initial setting of the shift register can be used as a key, but that only changes the starting point in the generated sequence, making shift registers relatively inflexible. Shift register systems are simple to implement but are only really suitable for low security applications.

3.4.6   Product Ciphers

3.4.6.1   Introduction

Both transposition and transformation ciphers have their advantages. A secure cryptosystem can therefore be constructed by combining both. A *product cipher* is a cipher where two or more cryptographic functions are concatenated – performed one after the other. The wartime Enigma cipher machine was an example of such a system, where a series of code wheels within the machine permuted and transformed message symbols into code symbols. Many commercial ciphers are based on the principle of performing relatively simple permutations and transformations enough times to form a secure system.

A good example of this type of system is the Data Encryption Standard (DES), which uses a sequence of 16 transformations and permutations. It has a 56 bit key, which is now susceptible to brute force attacks, but is still very popular. Other examples of product ciphers in commercial use are Blowfish, by Bruce Schneier, TC4, by Ronald Rivest (of RSA fame), and IDEA (International Data Encryption Algorithm), which uses a 128 bit key and a series of 8 transformations and permutations.

Product ciphers give a very good compromise between security, complexity, and key generation and distribution, and continue to be popular, being adopted by the US National Institute of Standards and Technology (NIST) for its replacement for DES, the Advanced Encryption Standard (AES).

3.4.6.2 DES

The Data Encryption Standard is a standard developed for the American government for use for financial transactions. It was developed from IBM's Lucifer cryptosystem. DES is a non-linear algorithm, which means that it is not possible to form valid messages by simple addition of cipher texts. DES involves only simple substitutions and additions, which makes it ideal for implementation on integrated circuits. Many such chips are now available, with processing speeds of the order of megabits per second. Another advantage of DES is that decryption uses the same equipment as encryption, only with the sub-key sections in reverse order.

As well as for encrypting transmitted data, DES is also used to encrypt passwords in the Unix operating system, and to check PIN numbers of ATM cash cards.

The DES algorithm consists of a series of 16 standard building blocks which permute and transform 64 bit input blocks into 64 bit outputs. Each Standard Building Block operates with a separate 48 bit key which is derived from the original key. The key has 64 bits, but 8 of these are parity bits, so the actual key is 56 bits long. Prior to being fed into the first Standard Building Block, the 64 bit input is permuted, and it undergoes an inverse permutation after the blocks (see Figure 3.10 (a)).

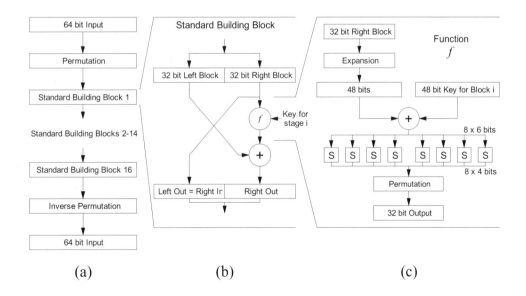

Figure 3.10   DES encryption

Each Standard Building Block transforms the left- and right-hand parts of the 64 bit input into a 64 bit output (see Figure 3.10 (b)). The right-hand part of the input is passed straight through to form the left-hand half of the output. The right-hand part of the output is formed from a non-linear function of the left and right inputs and the key for that particular block.

The non-linear function is detailed in Figure 3.10 (c). The 32 bit right-hand input is expanded to 48 bits by repeating half of its bits twice and permuting them. The 48 bit key for that particular block is then added. The result is divided into 8 6-bit blocks. Each of these blocks is used as an address for an S-box array. Each point in the array is a number from 0 to 15, so that the output is a 4-bit number. This S-box function therefore reduces the output to a 32-bit number. The S-box function is non-linear, i.e. $f(A) + f(B) \neq f(A + B)$, and each S-box is different. The 32 bit output of the S-boxes is permuted again before being added to the left hand 32 bit block to form the new 32 bit right-hand block output.

The 16 sub-key sections are formed from the 56 significant key bits by splitting the 56 non-parity bits of the key into two sections of 28 bits. These 28 bit sections are then shifted round by one or two bits and between each stage and at each stage 48 bits are extracted and permuted to form the sub-keys K_1 to K_{16}. If the sub-keys are used in the order K_1, K_2, $\ldots$, K_{16}, then encryption is performed. If the keys are used in the order K_{16}, K_{15}, $\ldots$, K_1 then the result is the inverse of the encryption function, i.e., an encrypted block is transformed back into its corresponding message block. This means that the same equipment can be used for encryption and decryption.

DES is very popular. The algorithm is published, and computer code to implement it is available in a number of different languages. The best attack remains an exhaustive search of all 2^{56} keys, although since the encryption of the complement of the message with the complement of the key yields the complement of the ciphertext, we can half the number of keys to try to 2^{55}.

Computer complexity is catching up with DES. Exhaustive search within a matter of hours has been possible for very rich organisations for a number of years and the advances in microchip fabrication now put such capabilities in the hands of anyone with a few tens of millions of dollars to spare. Encrypting messages twice leads to little increase in security due to the use of the so-called 'meet in the middle' attack. However, encrypting with one key, decrypting with a second, and re-encrypting with a third, which is called triple DES, increases the required number of keys to be checked to about 2^{80}, which can again be assumed to be safe.

3.4.6.3 Advanced Encryption Standard (AES)

As noted above, DES is being overtaken by the increasing complexity of processing hardware, which makes attacking DES a practical proposition. After an open invitation for algorithms which solicited 21 proposals from 11 countries, and a two year evaluation procedure, the US National Institute of Standards and Technology (NIST) has chosen a replacement for DES in the form of the Advanced Encryption Standard (AES). The cipher chosen was the Rijndael algorithm, a complex block product cipher which can be implemented efficiently. The Rijndael cipher (pronounced 'rain dahl' in English) was designed by Flemish researchers Joan Daemen and Vincent Rijmen. The standard defines three different key sizes: 128, 192 and 256 bits.

In a similar manner to DES, the algorithm consists of a number of 'rounds', the exact number of which depends on the size of the key, consisting of permutations and adding of a subkey. Although the Rijndael algorithm can work on other block sizes, the AES standard defines a block size of 128 bits. This block is split into 16 8-bit bytes, arranged in a 4 by 4 row and column arrangement. Each round consists of four stages: a non-linear S box transformation which can be implemented as a mapping from one byte to another, a permutation of the rows, a mixing of the columns, and finally an addition of the subkey for the current round.

One of the first applications of AES will be the encryption of information sent over the air in the new 3rd generation mobile radio system cdma2000.

3.4.7 Stream Ciphers

In a stream cipher, there is a feedback from the plaintext or, equivalently, the ciphertext, to the key. Using the message to form the key in this way is called *autokey*, and was first proposed by Vigenère in 1568. This has the advantage of reducing the length of the key which has to be stored or transported, but it has a very significant disadvantage in that, if there is any error in the message being sent, that error will propagate. In a block cipher, each block is considered separately so that errors will only affect a single block.

The Vigenère system is similar to the polyalphabetic system which now bears his name, but instead of repeating a keyword to define the alphabets, the message text is used after an initial key word (Vigenère used a single letter). This avoids the repetition which weakens polyalphabetic systems, but if one letter is corrupted, the decryption will be in error from that point on.

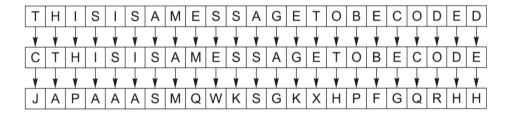

Figure 3.11 Encoding with Vigenère's autokey system

It order to decrypt the message, the receiver knows the key word or letter ('C' in the example shown in Figure 3.11), allowing decryption of the first letter of the message. This reveals the key for the next letter, and so on. Encryption and decryption are assisted by a Vigenère table, given in Table 3.1; the encryption of a letter in the first row by the alphabet given in the first column is the letter at the intersection of that row and column.

This autokey system is not confined to the Vigenère system, but can be used with other ciphers. The error propagation problem is serious, but using previous messages, either plain or enciphered, to affect the current encryption has the significant advantage that it breaks the one-to-one relationship between input blocks and output blocks that even complex block ciphers suffer from. For example, if DES system is used to protect a banking system, there might be a short 'transaction acknowledged' message which is the same in all cases. The eavesdropper could form a look-up table of ciphertexts without actually being able to decipher the messages. DES can be used in a streaming mode to prevent this, a popular one being 'cipher block chaining' (see Figure 3.12). The preceding ciphertext block is added to the message before it

Table 3.1 Vigenère Table

a	b	c	d	e	f	g	h	i	j	k	l	m	n	o	p	q	r	s	t	u	v	w	x	y	z
b	c	d	e	f	g	h	i	j	k	l	m	n	o	p	q	r	s	t	u	v	w	x	y	z	a
c	d	e	f	g	h	i	j	k	l	m	n	o	p	q	r	s	t	u	v	w	x	y	z	a	b
d	e	f	g	h	i	j	k	l	m	n	o	p	q	r	s	t	u	v	w	x	y	z	a	b	c
e	f	g	h	i	j	k	l	m	n	o	p	q	r	s	t	u	v	w	x	y	z	a	b	c	d
f	g	h	i	j	k	l	m	n	o	p	q	r	s	t	u	v	w	x	y	z	a	b	c	d	e
g	h	i	j	k	l	m	n	o	p	q	r	s	t	u	v	w	x	y	z	a	b	c	d	e	f
h	i	j	k	l	m	n	o	p	q	r	s	t	u	v	w	x	y	z	a	b	c	d	e	f	g
i	j	k	l	m	n	o	p	q	r	s	t	u	v	w	x	y	z	a	b	c	d	e	f	g	h
j	k	l	m	n	o	p	q	r	s	t	u	v	w	x	y	z	a	b	c	d	e	f	g	h	i
k	l	m	n	o	p	q	r	s	t	u	v	w	x	y	z	a	b	c	d	e	f	g	h	i	j
l	m	n	o	p	q	r	s	t	u	v	w	x	y	z	a	b	c	d	e	f	g	h	i	j	k
m	n	o	p	q	r	s	t	u	v	w	x	y	z	a	b	c	d	e	f	g	h	i	j	k	l
n	o	p	q	r	s	t	u	v	w	x	y	z	a	b	c	d	e	f	g	h	i	j	k	l	m
o	p	q	r	s	t	u	v	w	x	y	z	a	b	c	d	e	f	g	h	i	j	k	l	m	n
p	q	r	s	t	u	v	w	x	y	z	a	b	c	d	e	f	g	h	i	j	k	l	m	n	o
q	r	s	t	u	v	w	x	y	z	a	b	c	d	e	f	g	h	i	j	k	l	m	n	o	p
r	s	t	u	v	w	x	y	z	a	b	c	d	e	f	g	h	i	j	k	l	m	n	o	p	q
s	t	u	v	w	x	y	z	a	b	c	d	e	f	g	h	i	j	k	l	m	n	o	p	q	r
t	u	v	w	x	y	z	a	b	c	d	e	f	g	h	i	j	k	l	m	n	o	p	q	r	s
u	v	w	x	y	z	a	b	c	d	e	f	g	h	i	j	k	l	m	n	o	p	q	r	s	t
v	w	x	y	z	a	b	c	d	e	f	g	h	i	j	k	l	m	n	o	p	q	r	s	t	u
w	x	y	z	a	b	c	d	e	f	g	h	i	j	k	l	m	n	o	p	q	r	s	t	u	v
x	y	z	a	b	c	d	e	f	g	h	i	j	k	l	m	n	o	p	q	r	s	t	u	v	w
y	z	a	b	c	d	e	f	g	h	i	j	k	l	m	n	o	p	q	r	s	t	u	v	w	x
z	a	b	c	d	e	f	g	h	i	j	k	l	m	n	o	p	q	r	s	t	u	v	w	x	y

is encrypted, and an initial value (IV) is used for the first block. There is still a problem with very short messages which could still be recognisable, so the IV is often made dependent on the communication sequence number or some other variable known to both parties.

3.5   Public Key Cryptosystems

3.5.1   The Key Distribution Problem

Key distribution and management is a serious problem. In a private key cryptosystem, each communication link requires one key. For two users, only a single key is required, but from then on, each time a user is added to the system with $n - 1$ existing users, $n - 1$ additional keys are required. This means that a system with n users requires $\sum_{i=1}^{n-1} i$ different keys. In Figure 3.13, the system has 8 users, so 28 keys are required.

The public key cryptosystem has, as the name implies, a key which is public and therefore accessible to all parties who wish to communicate. This solves the key distribution problem – keys can be published in something like a phone book, for example. It also removes the need

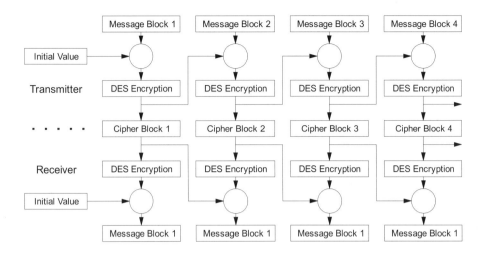

Figure 3.12 Cipher Block Chaining (CBC) operation of DES

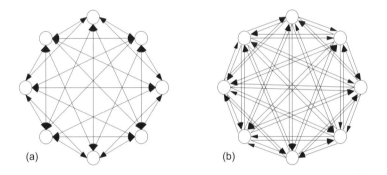

(a)   (b)

Figure 3.13 Keys for a private key system (a) and a public key system (b)

for communicating parties to agree a key beforehand. Alice can send an encrypted message to Bob even if she has never had any previous contact with him, which is important for such applications as electronic commerce, where secure communications with strangers is required.

However, a published key introduces a new problem. Once the message has been encrypted with the key, it should not then be possible to reconstruct the message from the enciphered text even with knowledge of the key, which, of course, is available to everyone. This requires the concept of a *one-way function*, a function which cannot (easily) be reversed, and a *trapdoor* so that with certain secret knowledge the recipient can reverse the process to obtain the message.

At first sight, it looks as if the public key system requires more keys, since each duplex communication requires two keys, one for each direction. However, in the public key system, although there are 56 communication links, every link with the same destination has the same key (the public key of that destination), so in fact only 8 keys are required. In addition, each communicating party only needs to keep one key – its private key – secret, instead of $n - 1$ keys in the private key case.

Public key systems are also called asymmetric systems because the process used for encoding is not simply reversed for decoding. This is a major difference from private key – symmetric – systems where the encrypting function is reversed for decryption.

3.5.2   One Way Functions

A one-way function is a function which is easy to calculate but difficult to reverse. An example is the multiplication of large numbers. We can calculate $89681 \times 96079 = 8616460799$ without too much difficulty, but the reverse process, the factorisation of the number 8616460799, is significantly harder. W S Jevons observed in 1873 of this sum that 'we can easily ... do a certain thing but may have much trouble in undoing it', an observation which is not only confined to arithmetic!

For the operation of a public key system such as the one described above we would like a function which when operated on the message, m, results in a cryptogram, c, from which it is not (practically) possible to discover m. In other words, we must be able to perform $f(m) = c$ easily (so that the cryptosystem can be implemented), but $f^{-1}(c) = m$ must be practically impossible. Note that reversal is always theoretically possible, even if only by calculating $f(m)$ for every possible message m until one matches c, but the cost of doing so must be greater than the value of the information which would be revealed, or the time taken to do so should be such that the information would be out of date.

One-way functions for cryptographic use require a further property – they must have a 'trap-door', a way by which someone with special knowledge can recover m from c with a reasonable amount of effort. This secret knowledge forms a *private key* which is kept secret by the recipient to allow decryption of the message.

Operation of the system is as follows:

- Each user has a pair of keys K_i and L_i
- Encrypt message m with $E(m, K_i) = c$
- Decrypt with $D(E(m, K_i), L_i) = m$
- Design E and D such that

    — given m and K_i it should be easy to find $c = E(m, K_i)$
    — given c it should be infeasible to find m (i.e. E should be one-way)
    — given c and L_i it should be easy to find m (i.e. E should have a trap door given by L_i)

Various functions have been proposed for use in cryptosystems:

- **Diffie-Hellman Key Exchange**: Based on the difficulty of finding discrete logarithms compared to raising to a power over a field. It is not really a cipher, but was the first published public key system.
- **RSA**: Based on the difficulty of factorising a product compared to multiplication. It is the most popular public key system, used for the PGP mail program, for example, but it is computationally expensive and commercial use in the USA is subject to patents.
- **Knapsack**: Based on the difficulty of separating a sum into its individual terms compared to adding the terms up in the first place. There have been a number of successful attacks on this system and, as a result, it is seldom used. However, it is much less complex than RSA, and until the advent of the more recent elliptic curve systems, it was seen as holding promise as a low complexity but relatively secure system.

- **Elliptic curves**: Elliptic curves are lines defined over a prime field. Given a point on the line, it is easier to use this point to calculate other points, but the operation is difficult to reverse without generating all possible points, which is not practical if the field is large. These schemes show great promise because encoding and decoding can be done at relatively low complexity, sharing the advantages of the knapsack schemes without their security concerns.
- **Error correcting codes**: As discussed in Section 5.4 on error control coding, while an error correcting code will be able to correct errors up to $(d-1)/2$, without a set of decoding rules the only way to do this is to compare the received vector with every codeword and choose the one which lies within distance $(d-1)/2$ of it. For a large code, this is practically impossible. It is therefore possible to form a cryptosystem by using an error correcting code and scrambling the generator matrix. The original generator forms the private key, allowing messages to be decoded, while the scrambled generator is the public key. Messages which are sent are disguised by adding errors, which an eavesdropper cannot remove, since the scrambled generator does not allow simple decoding.

3.5.3 Diffie-Hellman Key Exchange

The Diffie-Hellman Key Exchange protocol is not really a public key system in the normal sense since the transmitted secret is random, but it can be used to transmit information by using the shared secret to encrypt transmitted data. It was the first published public key system, and depends on the difficulty of finding logarithms over prime fields compared to that of calculating exponents.

The protocol exchanges a secret between two parties over an insecure channel, without requiring and existing secret knowledge. It works as follows:

1. Alice and Bob agree on a generator g and prime modulus p.
2. Alice generates a random number x. This is her private key. She calculates her public key X which is equal to $X = g^x \bmod p$.
3. In a similar manner, Bob generates a random private key y, and public key $Y = g^y \bmod p$.
4. Alice and Bob exchange their public keys X and Y.
5. Alice receives Y and uses her private key x to calculate $X^y \bmod p = g^{xy} \bmod p = k$.
6. Bob can also calculate k from the X he receives from Alice and his private key y since $X^y \bmod p = g^{xy} \bmod p = k$.
7. Both Alice and Bob now know k, but an eavesdropper cannot calculate k from observations of g, p, X, and Y.

As an example of such an exchange, consider the following example:

1. $P = 11$ and $g = 2$.
2. Alice gets random $x = 4$, and Bob chooses $y = 6$.
3. Alice calculates $X = 2^4 \bmod 11 = 5$.
4. Bob calculates $Y = 2^6 \bmod 11 = 9$.
5. Alice sends 5 to Bob, and Bob sends 9 to Alice.
6. Alice finds $Y^x \bmod 11 = 9^4 = 6561 = 5 \bmod 11$.
7. Bob finds $X^y \bmod 11 = 5^6 = 15625 = 5 \bmod 11$.
8. Therefore $k = 5$.

Note that the shared secret is random, since $k = g^{xy} \bmod p$, so it cannot be used to send information directly. If Alice and Bob do not choose their private keys truly randomly, this knowledge could be used to attack the system.

As P and g are published, attacking this system is effectively a discrete logarithm problem. $X = g^x \bmod P$ means $x = \log_g X \bmod P$. Therefore, Eve could find k by calculating $Y^{\log_g X} \bmod P$. In the above example, finding $\log_2 5 \bmod 11$ is trivial, but it is very hard if P is large.

3.5.4   RSA (Rivest, Shamir, Adleman) Cryptosystem

The RSA scheme depends on the relative difficulty of factorising a number compared to that of multiplication. It is one of the most popular systems, being used, for example by the PGP mail encryption program. It is patented in the United States.

In order to generate a key, two large random primes p and q are chosen. How large is a matter of debate and would depend on the application, but numbers in the order of a hundred digits are often used. These primes are kept secret, but their product $n = pq$ and a random e relatively prime to $(p-1)(q-1)$ are published and form the public key. The private key d is chosen such that $de = 1 \bmod (p-1)(q-1)$. Note that since p and q are kept secret, any eavesdropper cannot calculate d without factorising n.

Encryption of message m is by calculating $c = m^e \bmod n$. Decryption is by calculating $c^d \bmod n$, the result of which is m.

The fact that decryption works (i.e. that $(m^e)^d \bmod n = m \bmod n$) is not obvious, but depends on Euler's Theorem. This states that $a\phi(n) = 1 \pmod n$, for a relatively prime a and n with $0 < a < n$. $\phi(n)$ is Euler's totient function, the number of positive integers less than n which are relatively prime to n. If n is prime, $\phi(n) = n - 1$. If n is the product of two primes p and q, as is the case here, $\phi(n)$ is the number of numbers up to $n(n - 1 = pq - 1)$ less those which are multiples of p (of which there are $q - 1$) and those which are multiples of q (of which there are $p - 1$). No number less than n is a multiple of both p and q, so $\phi(n) = pq - 1 - (p - 1) - (q - 1) = pq - p - q + 1 = (p - 1)(q - 1)$.

This gives us that $de = 1 \bmod \phi(n)$ by definition. This means that $de = i\phi(n) + 1$ for some i. Decryption yields $c^d \bmod n = m^{ed} \bmod n = m^{i\phi(n)+1} \bmod n = (m^{\phi(n)})^k m \bmod n = (1)^k m \bmod n = m \bmod n$.

The RSA uses exponentiation, which is computationally intensive. For this reason, some schemes use RSA to set up a session key, and then use a faster, but less secure scheme for the actual transmission of data.

The obvious attack on the system – factorising n to find p and q – takes about $e^{\sqrt{\ln(n)\ln(\ln(n))}}$ operations based on the best factoring algorithms available. There are about 2^{25} seconds in a year, so at one billion operations per second, using a single processor it would take about 2^{41} years to factor a 1024 bit key (as used by PGP security algorithm). This is longer than the universe has been in existence since the big bang, and compares to about a year to break DES using similar assumptions. Using many computers in parallel brings the time to break DES down to only a few hours for organisations like governments, with a lot of money to throw at the problem, but the scale of the problem with RSA makes the task impossible without using another approach. Another attack would be to find x such that $x^e \bmod n = c$, but this would only yield that individual message and not break the system as a whole, and the problem is conjectured to be as hard as factoring in any case.

However, this excellent security can be completely undermined by poor choice of keys, and since the security of the system depends on the primes being large, there is a problem of finding suitable large prime numbers. There are a number of techniques for finding large primes which are suitable, but any system based on RSA must have a system for generating these numbers locally for key generation, since if the primes are communicated, or worse, taken from a list, security could be seriously compromised.

3.6   Authentication

3.6.1   Introduction

Communicating information secretly is only part of the problem. It is just as important to know that that information can be trusted – it has not been tampered with or received with errors, and that the person at the other end of the communication link is who they claim, and can be held to the message that they sent. These three additional aspects of security are message integrity, terminal authentication and message signatures respectively.

3.6.2   Integrity

Getting the information from one place to another in an encrypted form is one problem. However, in a communication system we must also be assured that it has arrived in an uncorrupted form. Corruption could be due to blocking of the communication link by an attacker, or by the attacker tampering with or even substituting the message, or from normal errors in the communication channel.

A simple way to ensure correct receipt is to ask for the message to be sent back. However, care is required for such protocols. If a private key system is used, the keys used by Alice and Bob will be the same, so all he has to do is send the encrypted message back to Alice, it will be the same as the cryptogram sent from Alice to Bob. If Eve can intercept the link, as opposed to simply tapping it, she can repeat the cryptogram back to Alice without decoding it. Alice would then think the Bob had received and understood a message from her when in fact it never reached him. Such message interception is more difficult than tapping in a conventional communication circuit, but on packet-based systems like the Internet, where messages are passed from machine to machine, it is trivial to discard or substitute messages by accessing an intermediate computer. Eve cannot decode the message but she can fool Alice into thinking that Bob got the message when in fact he did not. If the message is something like 'pay $1,000,000 into account X', such deleted messages could be very costly.

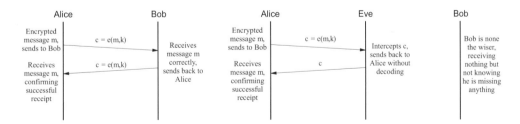

Figure 3.14   Attack on simple repetition acknowledgement protocol

One way to prevent this attack would be to encrypt the address and thus change the message being returned. However, care is required. If the encrypted address is sent as an addition to the message, we have $c = e(m, k) + e(a, k)$. Now all Eve has to do is listen to a few valid messages between Alice and Bob to find out $e(a, k)$, where a_A is Alice's address and a_B is Bob's address. When she wants to intercept a message, she strips $e(a_A, k)$ and replaces it with $e(a_B, k)$. Sending the time of receipt, for example, in addition to the address, may not help since Eve could calculate this and could replay part of an acknowledgement from a suitable time. Worse, such a protocol gives Eve a known plaintext attack, because she knows the address in plaintext and its encrypted form. She could use this knowledge to try to calculate the key k.

These attacks can be avoided by making the message a function of the information to be sent and the address, rather than appending the address to the message. We now have $c = e(f(m, a), k)$. Now even through Eve knows a_A and a_B, she cannot calculate the correct response $e(f(m, a_B), k)$ to fool Alice.

At first sight, public key systems are better than symmetric ciphers in this regard since encrypted messages from Alice to Bob, encoded with Bob's public key, are different from messages from Bob to Alice which use Alice's public key. However, poor protocol design in this case can lead to Eve deciphering the message rather than simply blocking it.

If Bob does not otherwise know the address Alice is transmitting from, all Eve has to do when passing the message on is to substitute her address for Alice's. Bob replies by encoding the message with the public key of 'Alice', which is actually that of Eve since she swapped addresses. Since the reply is encoded with Eve's public key, she can decode it and recover the message. She then re-encodes the message with Alice's public key and passes it on so that Alice is unaware of the interception. As in the symmetric cipher case, this attack can be prevented by making the message a function of the address and the information to be sent in such a way that the address cannot be extracted independently.

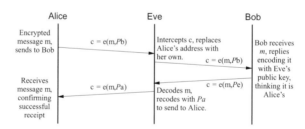

Figure 3.15   Substitution attack on addressed acknowledgement scheme

Rather than sending the message back to the source, some of these problems can be avoided by sending back a hash of the message, $h(m)$. A hash function maps an input set to a fixed length output. The hash function should be easy to calculate but very hard or impossible to invert, and should be such that $h(m_1) = h(m_2) \Rightarrow m_1 = m_2$ with high probability. The larger the fixed length of the output, the higher this probability is, but the more information has to be sent. One-way functions meet these requirements, so the public key of a public key system could be used, but it is not particularly easy to calculate. However, no private key is required. Since Alice knows m, she can encode it to find $e(m, Ph)$ and compare that to the received hash function. This is as difficult for Eve to attack as the original cryptogram sent by Alice. DES and MD5 are quite popular for hash functions, since they involve less computational complexity than the one-way functions used for public keys.

3.6.3 Authentication

In some cases, this trust in the authenticity of the communicating terminal is more important than the transmitted information itself. For example, in a mobile phone system, the actual communication itself may not be encrypted (although it often is), but it is important for the network to know that the communicating terminal is authorised to use the network and which terminal it is so that the subscriber can be billed for the call. The problem resolves to one of Bob needing to know that the person he is talking to is Alice and not Eve.

The FBI categorised authentication systems based on something a person knows (such as a password), something they have (like a key) and something they are (like a fingerprint or retina pattern). Any authentication scheme should then be based on at least two categories. This is appropriate for physical security, but still leaves the problem of remote communication. A system cannot depend simply on Alice saying 'I am Alice', because Eve could record a previous authentication session and replay the 'I am Alice' message. A good authentication protocol should involve a challenge followed by a response based upon the challenge (to avoid the replication attack) which proves the communicator is who he or she claims to be.

A public key system can provide a basis for authentication. Only the genuine communicator will know their own private key, and we can use this fact to authenticate them. In a public key system with encryption E and decryption D, $D(E(m)) = m$; the decryption of the encryption of message m yields the original message. However, if we start with a cryptogram, we can have $E(D(c)) = c$; the encryption of the decryption of cryptogram c yields the original cryptogram.

We can build a protocol around these facts. If Bob wishes to authenticate Alice, we proceed as follows:

1. Generate a random cryptogram, c, i.e., a message over the same alphabet as the enciphered messages.
2. Send the cryptogram to 'Alice' to decrypt.
3. 'Alice' 'decrypts' the message by using her private key. Since the original cryptogram was random, the decrypted message will be meaningless, but only someone who possesses the private key could generate it.
4. 'Alice' sends the decrypted message, m back to Bob, who then encrypts it with Alice's public key.
5. If the re-encrypted message matches the random cryptogram Bob originally generated, i.e., if $E(m) = c$, 'Alice' must have the private key and so must in fact be Alice.

If Eve was pretending to be Alice, she would have to be capable of generating an m such that if $E(m) = c$ for the c sent by Bob. This is equivalent to saying that Eve can decode cryptograms, and so would have broken the cipher. Some care is, however, required that the c generated by Bob is truly random. If it was not, Eve may be able to work out a suitable value for m for the particular value of c used by Bob, perhaps by observing previous authentication attempts. A Pay TV Smart Card system was defeated that way since the 'random' challenge was in fact an 8-bit binary number. Hackers defeated this system by recording all possible authentication challenges and the responses given by genuine cards, and then programmed a cloned card with a look-up table of valid responses without needing to know anything about the operation of the algorithm. This attack was possible because the genuine cards had extremely limited memory and processing power.

3.6.4   Digital Signatures

Alice and Bob may be confident of each other's identities, but that is not the same as trusting their word. It must be possible to attribute specific messages to their originator, so that others are able to verify its original and so that the originator can be held to their word as given in their document. This is done by *digital signatures* which fulfil the same purpose as their conventional counterpart, i.e., to 'sign' a message or receipt so that it can be proved that Alice sent it.

This is possible with a private key system if keys are kept by Alice, Bob and a registry. Bob will only accept messages as signed if he can decrypt the signature with his key. The central registry can check the signature in a dispute, and if it can also decrypt the signature then it will arbitrate in favour of Bob. However, this means that a central registry has to be established, and that registry will hold some information about Alice and Bob.

Public key systems give us a way out of this problem by using a similar process to that of authentication. In this case, we use the message or hash of the message as the 'random' cryptogram. Alice signs the document by using her private key to 'decrypt' the hash of the message. This decrypted message is the signature. Alice then sends the message and the signature to Bob, encrypting everything with Bob's public key for sending. Bob decrypts the message with his private key, and checks the signature encrypted with Alice's public key equals the hash of the message.

3.7   Other Cryptographic Protocols

Encryption and authentication of data are the two most obvious cryptographic applications, but there are a number of other cryptographic functions which do not fall into these categories which are becoming more important in communications but which cannot be solved by 'traditional' methods.

3.7.1   Tossing a Coin Remotely

The remote coin toss example is a good one for demonstrating the type of design required by good protocols. Consider the following scenario. Alice and Bob decide to go out on a date. Alice wants to meet at a restaurant but Bob would prefer a pub. When discussing the evening's arrangements by phone, Bob suggests he tosses a coin, and invites Alice to call. She says 'heads', whereupon Bob reports that unfortunately the coin came up tails, so they are off to the pub. Can Alice trust Bob? If fact, even if Bob reported that the coin came up heads, how is Alice to know if that was the true answer, or if Bob was simply being polite? A good protocol should protect against false losses as well as false wins.

Tossing a coin in the traditional manner works because the party tossing the coin cannot influence the outcome depending of the choice of the caller. If Alice calls and Bob tosses the coin, we require a protocol which forces Alice to choose an outcome before the toss, which Bob can hold her to after the toss, but from which Bob cannot gather which outcome she has chosen until after he reveals the outcome of the toss, so that he cannot cheat either. A suitable protocol can be based on the solution of roots over prime fields.

The equation $x^2 = a$ has two solutions (roots) for any positive square number a. For example, if a is 4, x can be 2 or -2. The same applies for fields of numbers taken modulo a prime p. $x^2 = 4 \bmod 5$ has solutions $x = 2$ and 3 ($= -2 \bmod 5$). (Note that a has to be a square number in that field. The above case has no solution for $a = 2$ or 3.)

If instead of using a prime modulus, we use a composite made up of two primes, we have 4 roots. For example, if $p = 3$ and $q = 5$, $u^2 \bmod 15 = 1$ has solutions 1, 4, 11 ($= -4 \bmod 15$), and 14 ($= -1 \bmod 15$). For many primes p, there are methods of calculating the roots mod p, but for large composite numbers this is very difficult. However, if the composite number is $n = pq$ and we know p and q, we can calculate the roots from the roots mod p and mod q. The protocol proceeds as follows:

1. Alice chooses two large primes p and q. She sends their product, $n = pq$ to Bob.
2. Bob randomly chooses an integer u between 1 and $n/2$ inclusive, and sends its square to Alice. This square has four roots, two of which will be between 1 and $n/2$ (since if x is between 1 and $n/2$, its negative, $-x$, won't be, and vice versa).
3. Using her knowledge of p and q, Alice computes the four square roots of $z \bmod n$. The roots are $\pm x$ and $\pm y$. She chooses the two between 1 and $n/2$. Let these be x' and y'. She now knows the u is one of these, but not which one.
4. Alice chooses one of these, and sends it to Bob. This is done by finding the smallest i such that bit i is different in x' and y' and a guess at whether bit i of u is 0 and 1. Note that Alice does not send the whole number to Bob, since if she had chosen incorrectly, Bob would then know both u (which he chose) and the other root (sent by Alice), and could then claim that Alice won when she did in fact lose.
5. Bob tells Alice whether her choice is correct or not, and sends her u to prove it.
6. Alice sends p and q to Bob so that he can find the other roots and verify that Alice followed the procedure.

Note that this protocol shares the secret so that at each stage neither party knows enough to cheat the other. Since there are two solutions to the root in the correct range, Alice does not know u, only that it can be one of two numbers. However, Bob cannot cheat by changing to the other solution after being told Alice's call because he himself does not know the other solution. This would require him to factor n which he cannot practically do.

3.7.2 Oblivious Transfer

The 50% chance of success evident in the remote coin tossing example can be used in protocols for 'oblivious transfer'. Oblivious transfer is where a secret is shared (by Alice, say) in such a way that she doesn't know (or is oblivious to) whether Bob has received the secret or not. While this might seem somewhat pointless in a communication system, there are cases when you wouldn't want the 'owner' of a secret to know whether the secret had been accessed or not. An example would be a voting system where it is important that ballots be verified but that the person casting them remains anonymous. The system proceeds as follows.

1. Alice chooses $n = pq$, where p and q are large primes. She sends n to Bob.
2. Bob chooses x and sends its square mod n to Alice. Note that unlike the coin tossing case this need not be the one between 1 and $\frac{n}{2}$. At this point Bob knows two of the four roots (x and $-x$).
3. Alice computes the square roots are $\pm x$ and $\pm y$, and sends one of them to Bob.

4. Since Alice does not know the number Bob chose, there is a 50% chance that Alice will send Bob one of the roots he already has. If Bob gets one of these, then he gains no information, and does not have the secret. However, if he gets the other root, then he can use this information to factorise n into p and q. Therefore with a probability $\frac{1}{2}$, Bob gets the secret, but Alice does not herself know whether Bob got the secret or not.

This basic system can be extended to the case where Alice gives one of two secrets to Bob, but doesn't know which secret Bob received.

The application of oblivious transfer does not seem obvious at first sight, but in fact it achieves a considerable degree of privacy for the information receiver. Normal information transfers are anything but private, with a trail of acknowledgements meaning that it is possible for an observer to know who knew what and when. However, there are times when privacy is required, for example, in electronic voting. It is possible to extend the type of protocols demonstrated in oblivious transfer to allow Bob to cast a vote with Alice, and for Alice to verify the vote itself as being valid, without being aware of how Bob cast his vote. This is essential for true electronic voting, for otherwise the government or election organiser would know which way each individual cast their vote.

These type of protocols also have application to electronic cash systems. Conventional cash has the following properties:

- **Secure from forgery**: It is not easy to mint cash, so forgery is relatively difficult.
- **Universal accreditation**: Everyone trusts cash, because it has the backing of a bank, and it is known everyone else will accept it.
- **Anonymity**: Possession of cash does not link you to a particular transaction, and those involved in a transaction do not need to know or trust each other to conduct the transaction. The trust is in the cash, not the person.

To replace hard cash, an electronic system would have to fulfil these requirements, and the anonymity requirement requires systems like oblivious transfer. However, many governments would like to electronic cash to be traceable to avoid making money laundering easier.

3.8   Practical Security

In this chapter, we have considered a number of ways for the communications engineer to build a cryptographically secure transmission system. Over the past few years, with the development of good cryptographic algorithms we have reached a stage where there is a toolset of protocols available for use with various levels of security. The engineer has only to see that the design work is undertaken properly to ensure the correct requirements in the first place.

Good security requires a holistic approach. It is no good double-locking and bolting the door of your house only to leave a side window open. There are a number of factors which do not directly relate to the communication system, but which are essential to ensure that the system is secure.

3.8.1   How Much Security?

An important question is how much security should be provided. It is now possible to use very secure schemes, but as systems become more secure, their complexity rises, increasing data transmission and processing costs. The cost of attacking the system also varies, as the more processing power used to attack the code, the faster an algorithm can be broken. There is a trade off between these three factors: the additional transmission and processing costs attributable to the security scheme, the cost of attacking the scheme, and the cost of the information itself (in terms of its value to an attacker or the loss to the owner). For example, an attack on a credit card transaction with a limit of £100 would not be worth throwing much processing power at, but to attack an inter-bank transfer system handling hundreds of millions of pounds would be worth investing in a few million pounds worth of computing power. For this reason, single DES is secure for the former but no longer for the latter, although triple DES would be. However, even single DES is too complex for small devices.

Data with a short shelf-life does not require as much protection. A system taking a matter of a week to decode would be perfectly suitable for sending battlefield movements, for example, because by the time the attacker decoded the data it would no longer be current. However, users would have to be aware of this and not use such systems for more long-term data.

Valuable data is not always obvious. The identity of a mobile phone customer may be considered to be much less important than the data they transmit, except if the fact that they are making a call from a foreign city alerts the attacker to the fact that their home is empty and can therefore be broken in to with impunity.

3.8.2   Trusted Authorities

Security protocols normally aim to require as little trust as possible, even, in the case of authentication and signature schemes, from the communicating parties themselves. However, some trust is always required, even if it is only in the individual user. Beware of non-obvious trust requirements. For example, using a security program for which the source code is unavailable means trusting the code writers (a) to know their job and write a secure program; and (b) not to cheat their users by including a hidden trap-door. Some users think that a system with a published source code must be less secure than one which is unpublished, but by Kerchoff's Assumption, the eavesdropper should always be assumed to have the code, and having the code published for the cryptographic community to see avoids both (a) and (b) above.

3.8.3   Error Conditions

In a number of ways, an undecipherable message is the same as an error condition. A well-designed communication system will cope with error conditions, so be careful that these procedures do not circumvent the security protocols. For example, a popular computer operating system had good security but a large loophole in that resetting the machine completely would boot the machine to a state where the user at the terminal could access, and change, any file. Early versions of another operating system would recover from the 'error' of a lack of a password file by allowing password-free access, so booting with a system disk and deleting the password files would allow access.

3.8.4   The Human Factor

The communications medium may be insecure, but it is dependably insecure. Often, the weak link in the chain is the human. Humans can circumvent systems in very innovative ways either maliciously or simply through lack of knowledge of the system. Poor choice of passwords or other key is an obvious flaw, but it is still one of the most common. Making the system more complex may seem to make it more secure but may involve the users circumventing it to save time or effort, for example by writing down keys which should be memorised. Finally, however secure the system, at some point a human has to operate it or interpret the data. That person could be bribed or coerced to circumvent the security.

3.9   Questions on the Security Perspective

1. Using a transposition cipher with a grid depth of 8, encode the message 'Takeover announcement will be made at noon tomorrow'.
2. Repeat the encoding using a transposition cipher with the keyword 'COLUMNAR' to define the ordering of the columns.
3. Encode 'Now is the time for all good men to come to the aid of the party' using Caesar cipher with a shift of 4.
4. Decrypt KWWS://ZZZ.DQWHQQD-PRGHOV.FRP/VHFUHW/LQGHA.KWPO.
5. Using the Vigenère cipher and the key word 'hide', encode 'We must meet under the clock.
6. Using the stream cipher and an initial key of 'D', encode 'The documents will be sent tomorrow.'
7. What is the bit stream generated by the following LFSR?

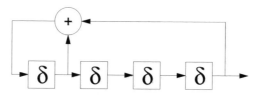

8. For an RSA system with primes $p = 7$ and $q = 13$, with encryption key $e = 5$, find the following.

   (a) What is the value of the private key d?
   (b) Encrypt the number 2 using your system, and verify that it is possible to decrypt the resulting codeword back to the message.

   Note that unless you are using a computer able to accurately work out the modulus of very large numbers, you will need to make use of the fact that

   $$a \times b \pmod{c} = (a \pmod{c}) \times (b \pmod{c}),$$

   and that
   $$a^{bc} = a^b a^c$$

4

The Network Perspective

4.1 Introduction

Until now we have considered the communication process in terms of a direct link between the sender and the receiver. While this is the view of the system taken by the user and the application, it is rarely the case in practice. Consider the system shown in Figure 4.1. In order to provide direct connections between all the various entities a large number of separate communication links are required. This is inefficient in a small system like this, and completely impractical for larger structures like telephone systems.

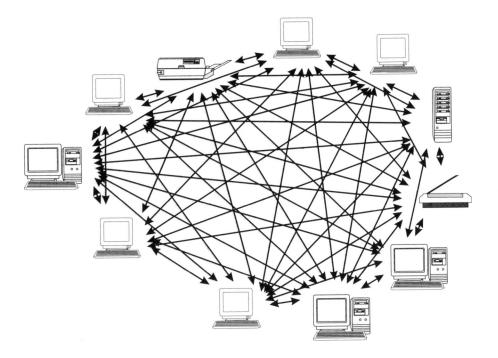

Figure 4.1 Direct connections between communicating devices

The answer is to use a network, which is a connection of communication links which can be switched to provide connectivity between terminals, as shown in Figure 4.2. In a network, the term *link* is used to refer to a communication path between two entities. These entities, which could either be terminals or intermediate switches, are known as *nodes*. Nodes without connections to the outside world, i.e. hosts or terminals, are called intermediate or switching nodes.

A network should have the following characteristics:

1. Delays, i.e. transmission and processing times, should be kept to a minimum.
2. It should make efficient use of its resources, i.e. bandwidth, and no part of the network should be idle for any extended period.
3. The cost of establishing, maintaining and operating the network should be kept to a minimum.

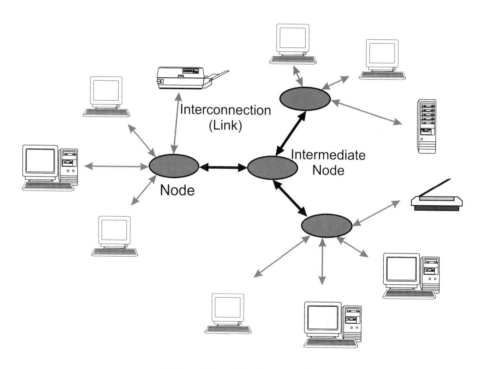

Figure 4.2   A simple network

4.2   Network Configurations

The simplest network is a pair of communication devices connected directly via some transmission media, i.e. in a *point-to-point* configuration, as shown in Figure 4.3.

The communication link can be anything from a simple pair of copper conductors to a radio channel to optical fibres. One can think of many examples varying from the trivial to the complex where such a configuration has significant merit; a simple household doorbell, an electronic radio-link baby monitor and RS-232 connections between PC and printer to name but a few.

Figure 4.3 Single point-to-point link

However, there will usually be more than two devices in the system. This leaves us with two options: a *switched* network or a *broadcast* network (see Figure 4.4).

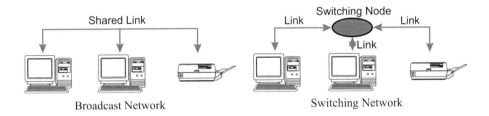

Figure 4.4 Broadcast and switching networks

In the case of the broadcast network, each of the terminals shares a common transmission medium. This means that they must take care to use the medium only when other terminals are not using it, and therefore in effect carry out their own switching. This gives rise to the concept of media access control (MAC) whereby some mechanism is put in place within each node to regulate how nodes transmit their data over the common medium. The term 'broadcast' comes from the fact that since the medium is shared, all messages are received by all terminals. Each terminal examines every message to see it is intended for it.

In the switching network, each terminal has a dedicated link to a switching node, which then connects it to the terminal it is communicating with, either directly or via other nodes. Unlike a broadcast network, data from a node in a switched network is only received by the node at the end of the channel and not by all others. Nodes in such a network are required to act as intermediaries and relay data or switch data on behalf of others in order that it might reach its destination. A routing function is implicit in such networks.

A convenient distinction can be drawn between *access networks* and *core networks*. The core network is the network connecting switching nodes. The access network is the part of the network which allows users to access the core network, and therefore consists of a network of terminal nodes and switching nodes. The distinction is relatively arbitrary and can be made on the basis of management convenience, so an access network may contain a network of switching nodes, although its primary aim will be to provide access to the system. In Figure 4.2, the grey links form the access network and the black links form the core network. The term 'core' leads to the concept of the access network forming the 'edge' of the network.

A distinction can be drawn between nodes within the access network and those in the core network. Core nodes are simply required to transport information between other core nodes and access nodes whereas nodes within the access network are likely to be terminals and are therefore also required to interact directly with the information sources. Access nodes can often be thought of as the generators or users of the information rather than solely as carriers of information; core nodes perform only the latter role.

A good example of the split is a telephone network. Here the nodes are the exchanges. Connections between exchanges are called 'trunks', and large numbers of calls are carried on these trunks. The access network is formed from a large number of individual subscriber lines feeding in to local exchanges. The core network has a number of intermediate exchanges to switch between local exchanges.

Core networks are usually switched networks since they are generally more efficient. Access networks are often broadcast networks because they are very easy to manage. Adding a node on a broadcast network can be done without affecting other nodes, so terminals can be added and removed easier. Broadcast networks are limited in size due to the shared link, so they are often formed of a number of pure broadcast sections with switches in between.

4.2.1   Network Topology

The topology of a network refers to the manner in which the nodes of that network are connected. There are four basic topologies: mesh, star, ring and bus.

4.2.1.1   Mesh Topology

A mesh topology is one in which the nodes of the network are directly connected to each other. In a full mesh all nodes are connected. In a partial mesh, some of these links are missing meaning that some nodes have to communicate through intermediate nodes (see Figure 4.5).

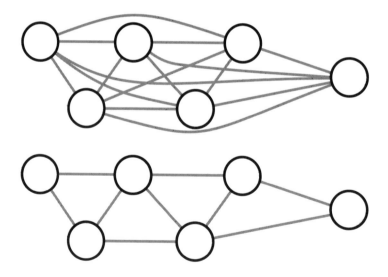

Figure 4.5   Full and partial mesh networks

With a full mesh, no switching technique is used as each node has dedicated links to all the rest, but we return to the highly impractical situation of Figure 4.1. The cost of adding new nodes to the network rises exponentially as the number of nodes increase, and even for small networks, the fact that every node must be altered when a new node is added is a significant deterrent. However, the central core network of some larger networks is sometimes arranged as a full mesh.

A partial network, on the other hand, is a practical and common network configuration. Nodes which have a large amount of traffic will be directly connected, while other nodes will be connected via intermediate nodes.

4.2.1.2   Star Topology

The star is a simple topology in which each node is connected by a direct link to a common central node (see Figure 4.6). This central node routes communications between nodes. The central node has to cope with the demands of all nodes and as a consequence it is both complex and expensive. There is also the problem of reliability, as if the central node fails, the whole network will fail along with it. To prevent such an occurrence there should be a high level of redundancy built into the central node through the provision of multiple processors, switches, etc. This provides a back-up should any part of the system fail. The PBX is an example of a star network designed to carry voice traffic.

4.2.1.3   Ring Topology

A ring topology, shown in Figure 4.6, is an arrangement by which all nodes in the network are joined by point-to-point links to form a closed loop. Information is then passed from node to node until it reaches its destination.

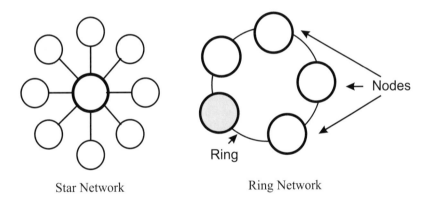

Star Network   Ring Network

Figure 4.6   Star and ring network

A ring network is an attempt at a more distributed network as the main communications processing power is no longer concentrated into one node. Each node has the ability to make decisions for itself regarding network data transfer and other functions. However, in a number of ring installations such as the Cambridge Ring, this distributed nature is reduced by the provision of a common node or *Monitor Station* which plays a central role in running the network. The Monitor Station monitors the network, collects error statistics, clears corrupted packets and performs the switch on/off functions for the network. Should this node fail, the operation of the network is likely to be seriously impaired. Another problem with this topology is the preservation of the closed loop; should the ring be broken in any way, then the network usually fails completely.

4.2.1.4   Bus Topology

This topology is an example of a fully distributed network. In a bus-based network there is
no central processing of any sort and the actual communications network is the transmission
medium itself. Each node simply attaches directly on to the cable or bus using the appropriate
interfaces and hardware. Both ends of the bus are attached to terminating devices, which helps
to provide the correct transmission characteristics for the bus.

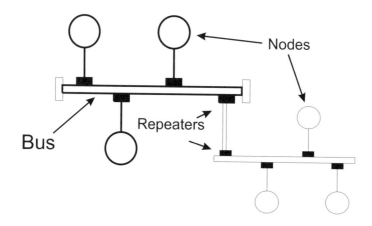

Figure 4.7   Bus and tree networks

This arrangement can be extended by using a number of buses, interconnected to form what
is sometimes called a *tree network*.

Both types of networks are shown in Figure 4.7 and unlike the ring topology there is no
requirement for a closed loop to be made by the transmission medium. As in the case of
the ring network, all stations share a common transmission link, so this means that only one
device can transmit at any one time. Thus some sort of access control is required to determine
when the stations can transmit onto the bus. The most common access method used in a bus-
based network is Carrier Sense Multiple Access (CSMA), with optional Collision Detection
(CSMA/CD) as used by Ethernet-style networks.

4.2.2   Connectivity

A key concept for a communication is one of connectivity – how easy it is to get from one
point in the network to another. In general, there will be a number of different ways to get
from one node to another. The specific choice of which way to go is the issue of routing.
However, at a higher level there is the issue of how well different parts of the network are
connected to each other, and so what routes are available. There are also reliability issues. If
a node or a link were to fail, would it still be possible for the network to operate?

While for very simple networks, it may be possible to ascertain their connectivity by
inspection, for networks in general the problem is more complex. For this reason we can
borrow from a mathematical concept called graph theory, because a network, a set of nodes
connected by links, is equivalent to a mathematical concept of a *graph*. *Directed* links have
traffic flow in one direction, whereas *undirected* links have two-way traffic flow. Nodes are

adjacent if they have a direct link between them. The *degree* of a node is the number of links ending on that node. The minimum degree of a network is the minimum degree of any node in the network. A *regular graph* has the same degree at all nodes. A ring is an example of a regular graph with degree 2.

A path is the sequence of nodes followed by traffic travelling from source node A to destination node B. The length of a path is the number of links in the path. The *geodesic* refers to the path between A and B with the smallest number of links. This is often called the shortest path but this is not strictly correct. The *diameter* of the graph (and therefore of the network) is the length of the longest geodesic.

Paths are *link disjoint* if they have no links in common. In Figure 4.8 paths ABCDE and AFGHE are link disjoint, while paths ABCDE and ABC are not. In the second diagram paths STUWX and SUX are link disjoint.

Paths are *node disjoint* if they have no nodes in common; the paths ABCDE and AFGHE are node disjoint while paths STUWX and SUX in the second network are not.

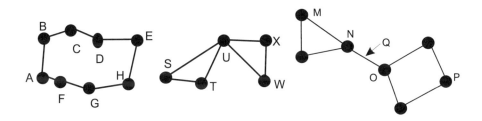

Figure 4.8 Link disjoint and node disjoint networks

The concept of disjoint paths is important when considering the reliability of a network. The two paths between nodes M and N in the third network are both node and link disjoint and should either a link or node in one of these paths fail, there is still an available path to transfer data between these two sources. On the other hand, there are two possible paths between nodes M and P but these are not link and node disjoint; should either node N or O or link Q fail, then no alternative path would exist and the connection would fail.

The reliability of a network – its ability to withstand node or link failure – is related to the number of node or link disjoint paths between source and destination nodes. The link connectivity between two nodes is the minimum number of links that must be removed to disconnect source from destination.

For example, the minimum number of links which must be removed to disconnect node C from node J in the network shown in Figure 4.9 is 3. This can be determined by drawing cut-sets across the graph as shown; the minimum cut-set is the cut-set which has the smallest number of links. The size of the minimum cut-set is the number of link disjoint paths between C and J, i.e. the link connectivity.

If we consider a different pair of nodes the link connectivity will be different. From node A to node L the minimum cut-set is line 4, giving a link connectivity of 2. The link connectivity of the whole network is the minimum connectivity of all node pairs. In a similar way, we can consider the node connectivity between each node pair, and the overall node connectivity is the minimum pairwise node connectivity. The link and node connectivities (C_n and C_l) and the minimum degree of the network(D) are related as $C_n \leq C_l \leq D_{min}$. In other words, the link connectivity of the network cannot exceed the minimum degree of the network and the node connectivity cannot be greater than the link connectivity.

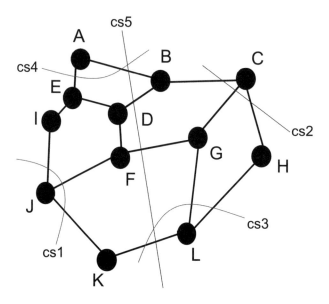

Figure 4.9   Different cut-sets for an example network

Since the degree of connectivity determines the relative robustness of a network, it is useful to be able to calculate the connectivity of a network. There are a number of algorithms which answer the question 'does this network have a connectivity of m?'. These algorithms test whether a network has a given connectivity; they do not determine what that connectivity is. However, it is possible to apply them iteratively to determine actual connectivity.

The suitability of the algorithm used will depend upon its speed in reaching a conclusion, and that will be related to the size of the network. Here, we will examine two such algorithms by Kleitman and Even.

4.2.2.1   Kleitman's Algorithm

Kleitman's Algorithm to test whether the network has a node connectivity of at least m proceeds as follows:

1. Choose any node N_1.
2. Verify that the node connectivity from N_1 to all other nodes is at least m.
3. Remove N_1 and all the links connected to it.
4. Choose a second node N_2.
5. Verify that N_2 is $m - 1$ connected to all other nodes.
6. Remove N_2 and all its links.
7. Choose a third link N_3.
8. Verify that it is at least $m - 2$ connected to all other nodes.
9. Repeat until reaching node m, i.e., choose node N_m, and check that it is at least 1-connected to all other nodes.

If all steps are satisfied, the network has a connectivity of at least m. However, if it fails at any point, the connectivity is not at least m.

As an example, consider the following network. Choose any node – B in this case – and confirm that $m = 3$ node disjoint paths exist between B and all other nodes in the network.

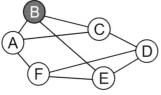

BA: BA, BCA,BEFA
BC: BC, BAC, BEDC
BD: BCD,BAFD, BED
BE: BE, BAFE, BCDE
BF: BAF, BCDF, BEF

This condition is satisfied, so remove node B from network along with its associated links. Choose another node – D – and verify that there are $m = 3 - 1 = 2$ node disjoint paths between D and all other nodes.

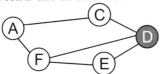

DA: DCA, DEFA
DC: DC, DFAC
DE: DE, DFE
DF: DF, DEF

This condition is satisfied, so remove node D from network along with its associated links. Choose another node – F – and verify that there are $m = 3 - 2 = 1$ paths between F and all other nodes.

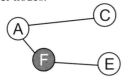

FA: FA
FC: FAC
FE: FE

Thus connectivity is at least 3.

4.2.2.2 Even's Algorithm

Even's Algorithm to test whether the network has a node connectivity of at least m proceeds as follows.

1. Number the nodes 1 to N.
2. Form any sub-set of nodes 1 to m, where m is the connectivity of interest.
3. Check that each node in this sub-set has at least m node disjoint paths to each other node within this group.
4. If the previous step fails then the connectivity is less than m. If successful, continue with next stage.
5. For each remaining node ($m <= j <= N$) form a sub-set of nodes (L) containing the set given in step 1) and incremental numbers of nodes given by the set J.
6. Add a new node X to the network and connect it to each node in set L. Verify that there are at least m node disjoint paths between X and each node j. Then add node j to set L, remove from set J and continue with next j.

If all the steps are satisfied, the network has a connectivity of at least m. Failure at any point means that connectivity is not at least m.

As an example, choose $m = 3$ nodes (B, E and F) and confirm that there are $m = 3$ paths between each pair of nodes in this set.

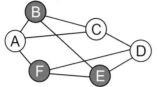

BE: BE, BAFE, BCDE
BF: BAF, BCDF, BEF
EF: EF, EDF, EBAF

Add a new dummy node X to network and connect to nodes B, E and F. Choose another node C and confirm that there are $m = 3$ paths between C and new node X.

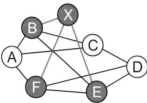

CX: CBX, CAFX, CDEX

Connect C to dummy node X. Choose another node A and confirm that there are $m = 3$ paths between A and new node X.

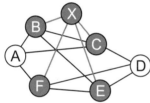

AX: ABX, ACX, AFX

Connect A to dummy node X. Choose another node D and confirm that there are $m = 3$ paths between D and new node X.

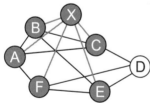

DX: DCX, DFX, DEX

Thus connectivity is at least 3.

4.3   Network Size

Networks can be classified in terms of size into:

- Wide Area Networks (WANs).
- Metropolitan Area Networks (MANs).
- Local Area Networks (LANs).

4.3.1   Wide Area Networks

A WAN is a data network which covers a very large geographical area (> 100 miles). These used to be formed either from leased lines from a Public Switched Telephone Network (PSTN) operator, but more recently deregulation in the telecommunications industry has led a number of telecomms and utility companies to set up high speed data networks on a country-wide or even global basis. The highest speed WANs operate at just under 10 gigabits per second.

4.3.2   Metropolitan Area Networks

MANs serve what is often called the 'metro area' and essentially provide data communications for areas of fairly high populations, such as cities. Originally the more limited operating distances of MANs (in the order of tens of kilometres) allowed MAN standards to define higher data rates than WANs, but more recent WAN standards allow speeds which are just as high, blurring the distinction between a MAN and a WAN.

4.3.3   Local Area Networks

A Local Area Network (LAN) is a facility of limited geographical range that links up a number of separate computing or data devices – i.e. non-voice services – contained within its site of operation, allowing the devices connected to the network to communicate with others on that site. Such a network can generally support a variety of bit-rates and services, and in terms of data delays and losses, offers relatively high performance characteristics.

Classifying data networks in terms of ownership would yield a similar breakdown to that obtained for geographical area. WANs tend to be owned by a network provider and the users of the network pay, either through rental costs of the leased lines, or directly, in relation to how much they used the network (like a telephone call). LANs are owned by the users and no charge is levied for their use. The main cost is associated with purchase and installation. In this respect MANs and LANs are similar, although a MAN may utilise a leased line provided by a network carrier.

4.4   Switching Techniques

There are two possible methods for arranging end-to-end communication through a network: connection-orientated and connectionless. In the former, a route is set up between the transmitter and the receiver for the information to be transmitted. A classic example of this is the telephone network. In the second case, the information is encapsulated into entities called *datagrams* which can make their way to the receiver through the network without a connection being established. The best analogy for this type of network is the postal service.

The distinction between the two types of switching is quite fuzzy. It is possible to implement a connection-orientation communication over a connectionless system. For example, to return to the postal analogy, if a company sent out daily dispatches to its managers by post, the end-to-end connection (company to manager) would be connection-orientated, because a regular communication path has been set up. At a lower level, the connectionless postal service implements the actual information transport, but this is invisible to the higher layers. One of the most common network protocols, TCP/IP uses this system. TCP is a connection-orientated protocol, while IP is connectionless.

Traditional telecommunications networks have been connection-orientated, while computer networks have been connectionless. Since computer to computer communications take place over telecommunications networks, often the classic connectionless inter-computer protocol IP is carried on a connection-orientated bearer, like ATM. This means that TCP/IP over ATM switches from connection-orientated at the application layer to connectionless at the network layer to connection-orientated again at the data link layer! The reason for this is the relative advantages of each approach at different layers in the system, as we shall discuss.

Another partition between switching systems is between *circuit switching* and *packet switching*. Circuit switching is equivalent to a connection-orientated system where a physical circuit joins the transmitter and receiver. Packet switching systems send data in packets. If the packets are treated individually by the network and routed to their destination, they are datagrams and the system is connectionless. However, the packets might not be self-contained but form part of a sequence. In this case they form a *virtual circuit*. The different switching classifications are shown in Figure 4.10.

Connection-Orientated		Connectionless
Circuit Switched	Packet Switched	
	Cell Switching	Message Switching
	Virtual Circuit	Datagram Switching

Figure 4.10   Relationship between different types of switching

4.4.1   Circuit Switching

In circuit switching, a dedicated communications path is set up between parties, via the nodes in the network. In traditional circuit switching architectures this connection remains in place for the entire duration of the call, although there are circuit switching techniques where this is not always true, i.e. fast circuit switching.

The advantage of circuit switching is that once a call has been set up, the users of that path can then transfer information unaffected by the load on the network, and this connection is guaranteed for the entire duration of that call. Transmission delays through the network remain constant and are usually negligible. The main delay component is the time required initially to set up the path through the network, the magnitude of which depends upon the actual network and path used.

The main disadvantage of this technique is its inefficiency. Keeping a path open for the entire duration of a call is extremely wasteful of channel capacity.

4.4.2   Packet Switching

If only a very small amount of information has to be sent, setting up a circuit to transfer the data is very inefficient. A better approach is to add address information to the data and send it in a self-contained unit, known as a packet, which is transmitted through the various nodes in the network to its destination. Packet switching has the advantage that the network is only used when there is information to transmit, whereas in circuit switching, the line is set up whether there is information flowing at that point in time or not. This is important for some services like data transfer which are very bursty.

The basic form of a packet is shown in Figure 4.12. The header, or overhead will contain information which will identify the source and destination of the packet as well as

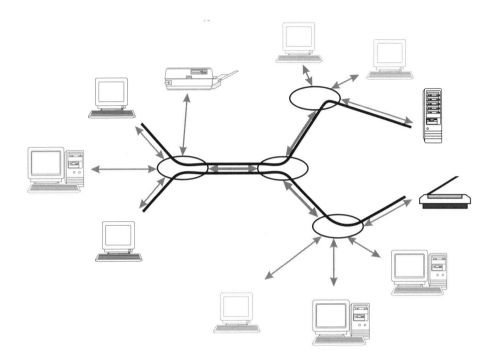

Figure 4.11   Circuit switching

Flag	Address	Information	Flag

Figure 4.12   Data packet

synchronisation bits which indicate the start or end of the packet. The packet is transmitted through the network stage by stage, being stored in a buffer until the associated stage in the path is free. There is additional information attached to the packet in order to identify its source, destination or route through the network, but this would be fairly small compared to the message itself.

There are two special types of packet switched systems. The first is *message switching*, shown in Figure 4.13, where the entire message forms a packet. Message switched systems are often referred to as store and forward systems. While message switching, avoids splitting the information to be transmitted into smaller packets, and the consequent reassembly at the receiver, there are a number of disadvantages associated with using message switching in a communications network.

1. Message sizes will vary from tens of bytes to tens of thousands of bytes and the hardware in the system must be able to store all messages. This is obviously inefficient and costly in terms of network resources.
2. Unacceptable delays could result with long messages which could also affect the delays associated with short messages.

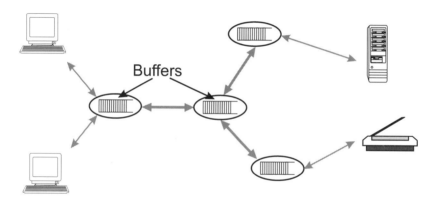

Figure 4.13   Message switching

The second special type of packet switching is *cell switching* (see Figure 4.14). In a cell switching system, all the packets, known as *cells*, have a fixed length. This common format reduces the amount of work the network nodes have to perform on the packet, keeps complexity low and speeds high. However, there is a trade-off in terms of the cell length. Short cells reduce delay, because each cell is read in to a node before being processed and sent out. However, if the cell is too small, the overhead in terms of the header will be a significant proportion of the data flow, which is inefficient. On the other hand, a large cell would increase delay, and cause difficulties with low data rate services like speech since much of the cell would go unused, again causing inefficiencies. ATM uses cell switching with a 48 byte information field and 5 bytes of header.

To keep the processing and header overheads small, cell switching systems use virtual circuits so that the path through each router is set up in advance.

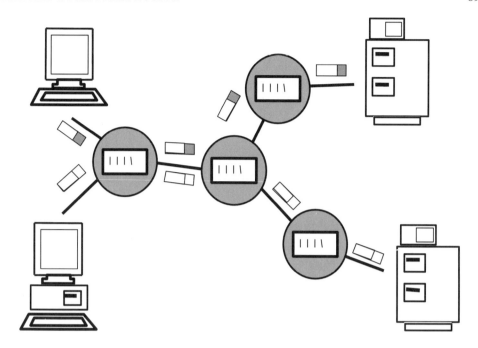

Figure 4.14 Cell switching

4.4.2.1 Datagram Switching

In a datagram-based system all packets are considered as separate entities and are routed through the system as such. This means that packets from the source, bound for the same destination, will travel by different routes and could possibly arrive out of sequence.

The route taken by the packet will depend upon the state of the network at that particular instant. A sequence number in the packet header will allow the correct sequence of packets to be established. Since each packet is routed through the network independently, the header must also contain source and destination addresses and possibly routing information.

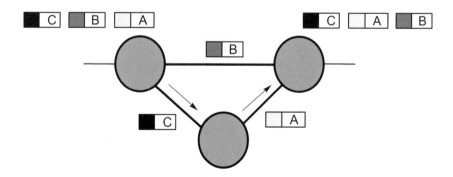

Figure 4.15 Data flow using datagram switching

4.4.2.2   Virtual Circuit

Virtual circuit operation is similar in some ways to circuit switching; a path is established between the source and destination pair (by the first packet) and the same path is followed by all subsequent packets from the same source. Consequently the packets cannot arrive out of sequence. This also allows an assignment of resources to the information flow for the session, since nodes know what to expect. In principle, the virtual circuit header is simpler than that of the datagram because there is no routing change between packets, and no re-sequencing is required.

A virtual circuit is a logical point-to-point path between two end stations. Normally, service has three phases: *call/circuit set-up*, *data transfer*, and close down (or *call clearing*). Full source/destination addresses are used to setup the virtual circuit. The virtual circuit is then assigned a virtual circuit number used by all subsequent messages. Normally error and flow control is provided over virtual circuit.

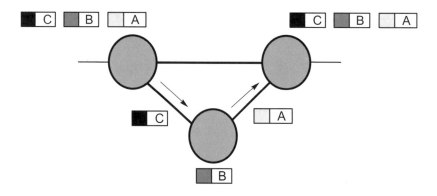

Figure 4.16   Data flow in a virtual circuit

The virtual circuit header is shorter than that of the datagram and thus, in principle, transmission efficiency is better. Delays can be reduced in a virtual circuit system as packets can be processed quicker by each node and re-sequencing is not required at the destination. However, these disadvantages are offset by the fact that it is easier to facilitate a communications service that crosses network boundaries using datagrams than virtual circuits.

4.5   Service Definition from a Network Viewpoint

The purpose of the communication system is to carry information between two or more points to the user's satisfaction. The user interfaces with the application. However, the network interacts with the data several layers below that, after the data has been processed in various ways. The most significant process, both in terms of altering the data and the user's perception of that data, is likely to be source coding.

This processing has a number of advantages for the network designer. While to the application, the service may be speech, video, or file transfer, to the network there is simply the requirement to transfer some data to the destination with some quality criteria. The ITU-T has defined a number of Quality of Service (QoS) criteria which it splits between *performance-oriented* and *non-performance-oriented* groups.

4.5.1   Non-performance-oriented Quality Parameters

Non-performance-oriented parameters do not directly affect the performance of the communications but are concerned with related matters.

- **Level of service**: This is the degree of certainty that the agreed QoS levels will be delivered. This can be defined as *deterministic*, where a given QoS level is guaranteed, *predictive*, where the service may suffer from QoS degradation from time to time because of the statistical nature of the network communications, or *best effort*, where the service only receives whatever network resources are available after other levels have been serviced and nothing is guaranteed.
- **Cost**: The charge for using the service and receiving the agreed quality.
- **Priority**: The precedence given to the service. High-priority connections are serviced before lower ones.

4.5.2   Performance-oriented Quality Parameters

Both the ITU and the ISO specify parameters for measuring the direct quality of a connection. The terminology is slightly different but generally equivalent. Where two terms are given below, the first is used by the ISO, and the second by the ITU,

- **Establishment delay, mean access delay or call set-up delay**: The delay between the issuing of a new call connection request and the confirmation that the connection has been established.
- **Establishment failure probability, probability of blocking**: The probability that a requested connection is not established (within the maximum acceptable establishment delay).
- **Throughput**: The maximum amount of data that may be successfully transferred per unit of time over the connection on a sustained basis.
- **Transit delay, frame delay**: The amount of time between issuing a PDU and its reception.
- **Delay variation, frame jitter**: Variance between the minimum and maximum transit or frame delay.
- **Residual error rate, Frame Error Rate (FER)**: The probability that a frame or PDU is corrupted, lost or duplicated at the receiver.
- **Resilience, probability of dropping**: The probability that the connection will be terminated or reset within a specified interval of time.
- **Release delay**: The delay between the issuing of an end of call request and the confirmation that the connection has been released.

The establishment delay and release delay are governed by signalling and routing within the network. While it is obviously useful to minimise these delays, these only occur once per connection.

Network availability is an important issue and is covered by blocking, not allowing a connection to be established, or *dropping*, terminating an existing connection. The latter is generally considered to be very unsatisfactory, it being preferable to block a connection in the first place rather than terminate it when it is in progress. The term *Grade of Service* is sometimes used to refer to the availability of the network in the first place, i.e. the establishment failure probability, to distinguish it from the quality of service when a service is actually delivered.

The three fundamental quality parameters for data transport are delay, throughput and corruption, which includes both errors and omissions. For each of these parameters, it is likely that their variation as much as their average value will be important. The variation in throughput or bit rate is referred to as *burstiness*, and is often defined in terms of the difference between the mean bit rate and the peak bit rate, with a service specifying its requirements for both these values. The variation in delay is referred to as *jitter*.

4.6   Network Dimensioning

Having defined the requirements on the network in terms of services, it is then possible to calculate the capacity we require for the network. There are large differences between the approaches required for circuit-switched and packet-switched networks.

4.6.1   Dimensioning in Circuit Switching

In a circuit-switched system, each terminal is provided with a dedicated circuit to the network. However, since it would be far too expensive to provide a full mesh network between terminals, the connections between nodes will not be able to support all the terminals at once. This is quite satisfactory, as most terminals will only be in use for short periods and so we only need to provide enough capacity on the core network for the expected maximum load. This is particularly true when there are a large number of terminals each used relatively infrequently, as is the case with a telephone system. How many resources do we require in order to satisfy the users, and how likely is it that a user will find all the resources occupied? We assume that a user, finding no resources available, does not wait but leaves the system. This is the case in a communication system where the resources are circuits.

If the population of users is not very much larger than the number of circuits, then the probability of a new user arriving will decrease as the number who currently hold a circuit increases. However, as long as the number of users is large enough, we can assume that the number who are currently using resources does not affect the probability of a user arriving in the system wanting a circuit. Let this probability, the call arrival rate, be λ. The system can be in a number of states, depending on the number of circuits in use. It there are n circuits, there are $n + 1$ states (since there is also a possibility that no circuits are in use). Let these states be denoted $E_0, E_1, \ldots, E_n$, and let the probability of being in state i be P_i. This means that the proportion of time that the system is in state E_i is P_i, so the probability of getting a call and moving to state E_{i+1} is λP_i, except for $i = n$, where the probability is 0 (since there are no more circuits).

The probability of a call finishing in state E_i depends on the length of the calls. Let the mean call length, called the mean call holding time, by τ. This means that the probability that a call will leave an individual circuit is $1/\tau$, so if there are i calls, the probability of any one of them leaving is i/τ. This means that the probability of a call leaving and moving to state E_i is $\frac{i+1}{\tau} P_{i+1}$. We have the situation shown in Figure 4.17.

Assuming that the system is stable (and all practical systems will have to be), the probability of entering a state and the probability of leaving a state must be equal. This gives us

$$\lambda P_i = \frac{i+1}{\tau} P_{i+1} \quad \forall\, i \in \{0, \ldots, n-1\}.$$

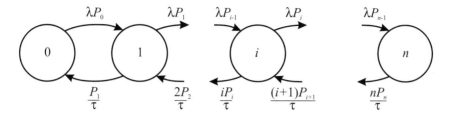

Figure 4.17 Different call states

We can express each of these probabilities in terms of P_0 as follows

$$P_i = \frac{(\lambda\tau)^i}{i!} P_0 \quad \forall\, i \in \{0, \ldots, n\} \tag{4.1}$$

However, the system must always be in one of the states, so $\sum_{i=0}^{n} P_i = 1$. Using this allows us to solve for P_0 to give

$$P_0 = \left(\sum_{i=0}^{n} \frac{(\lambda\tau)^i}{i!} \right)^{-1} \tag{4.2}$$

Substituting 4.2 in 4.1 gives the proportion of time a system will be in a state with j calls busy.

$$P_j = \sum_{i=0}^{n} \frac{i!(\lambda\tau)^j}{j!(\lambda\tau)^i} \tag{4.3}$$

The various proportions of time the system spends in each state depends on the product $\lambda\tau$ rather than either of these terms individually. The product of the call arrival rate and the call holding time gives the average load on the system. This quantity is called the *offered load* and is given the symbol A. It is measured in Erlangs after A.K. Erlang, who published the formula in 1917.

Erlangs are measured in calls. A point to watch is that call arrivals are often quoted as a number of calls per hour, since systems are dimensioned on the basis of the hour with most calls during the day (the *busy hour*). Call duration, on the other hand, is often quoted in minutes or seconds, and the two time intervals must be equal before calculating the load.

If we substitute $A = \lambda\tau$ into 4.3, and setting $j = n$, we get the *Erlang loss formula*.

$$E(A, n) = \sum_{i=0}^{n} \frac{i!A^n}{n!A^i} \tag{4.4}$$

The Erlang loss formula gives the proportion of time the system will be fully occupied (and so any arriving call will be blocked), and since an underlying assumption of the argument is that call arrival probability is independent of the number of calls in the system, this proportion will give the blocking probability of the system. It is normal to have a low value for blocking probability, of perhaps a few percent. The formula is often called the Erlang B formula for this reason (B for blocked). (There is an Erlang C formula which covers the case where calls arriving at a full system are queued.)

The foregoing discussion assumes that any circuit coming it to a switch can be connected to any free circuit out of the switch. This is easy to arrange with a device called a crossbar switch, where each input line may be connected to any output line via the relevant interconnection. Since a large proportion of the interconnections will not be used, the amount of hardware in this system may be reduced by a using a multi-stage arrangement. This operates similarly to two exchanges with a trunk in between, and may lead to blocking in the switch.

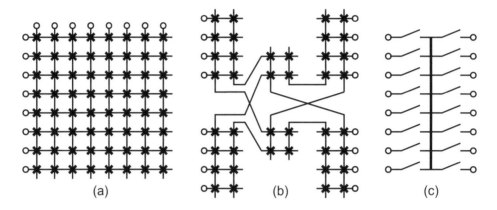

Figure 4.18   Different switch architectures (a) crosspoint, (b) multi-stage crosspoint, and (c) time division switch using a bus

Since the data provided to modern telephone exchanges is all digital, it is possible to take a different approach and use time division switching. Each digital line into the switch is connected to a common bus through a switch. The line is buffered, so that data can be read in and out at different rates. To switch between two lines, both are connected to the common bus, so the data can move between the buffers. This data is moved at high speed so that many separate time intervals are available, each of which can be used to send data between a pair of lines. If data was moved in the switch eight times faster than it arrived or departed on any line, then any eight pairs of lines could be connected. By running the switch fast enough, it is possible to make the switch non-blocking for a given number of lines, but the faster the switch, the more expensive it is to build.

It is possible to share the connection to the network between a number of terminals, which requires a switch to connect one of the terminals to the circuit and disconnect the others. A simple example is a telephone extension, where all the telephones share the same line and only one call is in progress at any one time. A party line is a similar arrangement between a number of subscribers. A slightly more sophisticated system is an automatic phone/fax and perhaps computer switch which automatically answers an incoming call to the correct device depending on whether it hears a fax or modem tone (connecting to the phone or answering machine if neither are heard). These systems work quite effectively for voice or fax calls since these calls are usually short. However, the same cannot be said for data (i.e. modem) calls, which can lead to familial disharmony and a good market for telephone companies in providing second lines to homes. Since a telephone line is physically capable of carrying the traffic of more than one call, one possibility is to digitise the signals of two calls and send them down the same line. This is the approach taken by ISDN, but it requires a more expensive phone or conversion equipment to take the usual analogue telephone signal and digitise it. (The equipment at the exchange does not need to be changed, since it is already digital.) A simpler solution is to provide a second analogue line. Since telephone cabling is usually provided with two pairs of wires, this is often just a matter of connecting up the second pair of wires.

4.6.2 Dimensioning in Packet-Switched Systems

Data traffic is, by nature, bursty. By this we mean that the traffic consists of (relatively) short bursts of activity followed by longer periods of inactivity. Contrast this with voice traffic, where information exchange once calls are setup is just continuous stream of data at a single rate. As result, data traffic is not ideally suited to being carried over its own dedicated, fixed capacity channel as in circuit switching. Such an approach can result in highly inefficient use of the channel.

Perhaps a better approach is to aggregate all the capacities assigned to individual data circuits and to share this among all the data users. The peaks and troughs of individual circuits can 'cancel' each other out with the net result that bandwidth that was originally assigned is now free to be used to support additional calls.

Statistical multiplexing facilitates more efficient use of system resources but must be used carefully. A key question is how many channels can be multiplexed together. Too few and the bandwidth gain is poor; too many and the bandwidth allocated is insufficient, resulting in increased delay or loss.

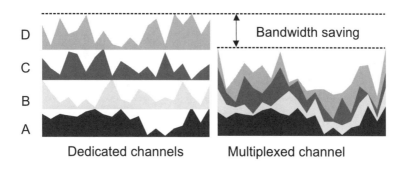

Figure 4.19   Statistical multiplexing

Using multiplexing procedures, it is possible to carry several separate data streams independently over the same physical medium. It is possible to use the physical properties of the medium to carry out multiplexing. This must be done at the physical layer and is covered in Section 6.5. Modern optical and radio networks in particular make use of this form of multiplexing.

The network layer has the option of time division multiplexing (TDM). TDM is simply the allocation of the transmission link to different services or users at different times. It is inherent in a packet based system, which is an example of asynchronous time division multiplexing (ATD). There is another type of TDM – synchronous time division multiplexing (STD), which can be used to implement circuit switching. In synchronous TDM, transmission frames are defined, each comprising a certain number of fixed-length timeslots. Specific timeslots are then allocated to a user, during which time the user may send or receive information units, and the position of a certain timeslot within a transmission frame identifies the channel (user). It is called synchronous because a certain timeslot always arrives at the same time during frame transmission.

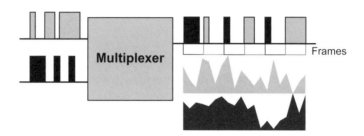

Figure 4.20   Synchronous time division multiplexing

The calculation of the capacity of a packet switched system which uses synchronous time division multiplexing is similar to the case of circuit switching. Each time division can serve one transmission, and so forms a virtual circuit. The number of circuits is therefore the number of time divisions. The situation can be complicated by the fact that not all the time divisions need be the same size, but can be adjusted to match the data requirements of the given virtual circuit. However, although the number of circuits may therefore change, the same basic approach still applies.

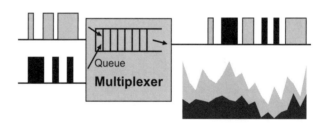

Figure 4.21   Asynchronous time division multiplexing

In an asynchronous time division system, there is no set allocation for each transmission path. Instead, packets arriving at the multiplexer are sent on as soon as possible. While waiting for onward transmission, packets are held in a queue. As the load on the system increases, so will the amount of time packets must wait in this queue for transmission. The capacity of the system will therefore be dictated by the maximum time users are willing to wait. This is a 'soft limit', since it is not constrained to a specific value. Circuit-switched systems have a 'hard limit' in the number of circuits available.

A queue is defined by four parameters, usually denoted $A/B/a/b$ – the distribution of arrivals, A, the distribution of served items (i.e. departures), B, the number of servers, a, and optionally, the maximum number of items which may be present in the queue, b. In the simplest case, a communication system can be considered to have arrivals at random. Such a distribution is represented by an exponential or *memoryless* distribution, denoted M. Other distributions are D, for deterministic, or E, for Erlangian, for example. If the packets are of variable size, then the serving distribution will also be memoryless. For simplicity, consider the queue to be of infinite size, so that no packets are ever lost because they encounter a full queue. Since there is one communication path from the multiplexer, there is one server, so the result is an $M/M/1$ queue.

To find the time taken for a packet to go through the multiplexer, we start with the state transition diagram, as shown in Figure 4.22.

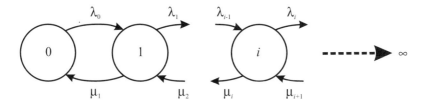

Figure 4.22 State transition diagram for an $M/M/1$ queue

As in the case of the Erlang discussion in Section 4.6.1, we can assume the system is stable, so the probability of moving up to a state must be equal to the probability of moving down from that state again.

$$\lambda_i p_i = \mu_{i+1} p_{i+1}$$

Unlike the Erlang discussion, where each call had its own circuit, here there is only one server, so departure rate μ will be constant and independent of the state of the system. We have also assumed that the input distribution is memoryless, so the arrival rate λ will also be constant and independent of the system state. Therefore, $\lambda p_i = \mu p_{i+1}$. $\lambda p_0 = \mu p_1$, and $\lambda p_1 = \mu p_2$, so $p_2 = \frac{\lambda}{\mu} p_1 = \left(\frac{\lambda}{\mu}\right)^2 p_0$. Generally, $p_i = \left(\frac{\lambda}{\mu}\right)^i p_0$, or $p_i = \rho^i p_0$, where $\rho = \frac{\lambda}{\mu}$. ρ must be less than 1 otherwise the queue would increase indefinitely.

Since the system must be in one of its states:

$$\sum_{i=0}^{\infty} p_i = \sum_{i=0}^{\infty} \rho^i p_0 = 1.$$

Using the fact that the sum to infinity of a geometric series ar^n (with $r < 1$) is $\frac{a}{1-r}$, this means that,

$$p_0 = \frac{1}{\sum_{i=0}^{\infty} \rho^i} = \frac{1}{1/(1-\rho)} = 1 - \rho$$

$$p_i = \rho^i p_0 = \rho^i(1 - \rho)$$

Utilisation is the proportion of time a system is active. In this system, it corresponds to the time there is something in the queue, i.e., $1 - p_0$. Since $p_0 = 1 - \rho$, the utilisation of the system is ρ.

We wish to find the mean delay to packets arriving in the system. To do that we first have to find the mean number of packets in the system, N. The probability that there are i packets in the system is given by p_i, so

$$
\begin{aligned}
N &= \sum_{i=0}^{\infty} i p_i \\
&= \sum_{i=0}^{\infty} i \rho^i (1 - \rho) \\
&= (i - \rho) \sum_{i=0}^{\infty} i \rho^i \\
&= (1 - \rho)(\rho + 2\rho^2 + 3\rho^3 + \ldots) \\
&= (1 - \rho)\rho(1 + 2\rho + 3\rho^2 + \ldots) \\
&= (1 - \rho)\rho \frac{1}{(1-\rho)^2} = \frac{\rho}{1-\rho}
\end{aligned}
$$

since $1 + 2x + 3x^2 + \ldots$ is the power series expansion of $\frac{1}{(1-x)^2}$.

Little's Law states that given items arriving at an average rate of λ, and staying in the system for an average time of T, the number in the system will average $N = \lambda T$. Since $N = \frac{\rho}{1-\rho}$, we have $T = \frac{N}{\lambda} = \frac{\rho}{\lambda(1-\rho)}$, or, since $\rho = \frac{\lambda}{\mu}$,

$$
T = \frac{1}{\mu - \lambda}.
$$

The assumptions regarding random arrivals and departures are unlikely to be valid in practice, but this analysis provides a useful starting point for real systems. In practice, data traffic arrivals are not truly random but exhibit what is known as long-range dependence – underlying trends which mean that groups of large packets occur together. This means that estimates based on memoryless queues overestimate the capacity of the system slightly. Detailed capacity planning of real systems is usually carried out by computer simulations.

Consider a link with a capacity of 20kbit/s. The average traffic is 14 packets per second, and the average packet size is 800 bits. The capacity of the link is therefore 25 packets per second, so the delay on this link is $T = 1/(25 - 14) = 91$ ms.

The equation given above $T = \frac{1}{\mu - \lambda}$ provides the basis to perform simple packet-based network dimensioning and the following example illustrates how it can be used. Consider the following situation; a packet based multiplexor provides, via an outgoing 64 kbit/s transmission line, connectivity to a data network for a number of user terminals. At the busiest period it is found that each user terminal generates, on average, 3 packets per second and the average packet length is measured as 400 bytes. The question posed is: 'How many terminals can be connected to the multiplexor?' To answer this question, it is necessary to relate the delay equation given previously into terms that relate directly to the packet network, providing we make some simplifications and important assumptions about the behaviour of the network.

- **Packet Delay**: The total delay experienced by an asynchronously multiplexed packet is made up of a number of key components – the propagation delay, the processing delay within the multiplexor, the waiting time (in the buffer) before the packet is processed and time taken to transmit the packet down the link. To the simplify the problem, we will ignore the propagation and processing delays; not an unreasonable step given that the both delays will be negligible (assume short links and fast processors) compared to the time taken to transmit a packet. The transmission delay is simply the packet length divided by the bit rate ($\frac{3200 \text{bit}}{64000 \text{bit}/\text{s}} = 50$ ms)
- **Service Rate**: The assumption that the total delay consists simply of the queuing delay and transmission delay allows the multiplexor to be represented as a simple queue if we think of the transmission delay as the packet holding time. As the packet service rate is the reciprocal of the mean packet transmission delay, the numerical vale is trivial to determine, $\mu = 64000/3200 = 20$ packets per second. In order to fit this parameter to the M/M/1

delay equation we will have to assume that packet lengths are random but follow a negative exponential distribution. If the transmission time and hence packet holding time follows a negative exponential distribution, then the service process can be described as Possionian and the M/M/1 may be applied. In reality, packet lengths will not exhibit such behaviour but the approximation is sufficient for a first order conservative estimate.

- **Arrival Rate**: Numerically each source generates 3 packets per second and we will assume that the inter-arrival time follows a negative exponential distribution in order to describe each packet source as Poissonian. Such an assumption is essential for two reasons: firstly to allow the M/M/1 delay equation to be applied and secondly to allow the total arrival rate be represented as the sum of the individual rates from each source. Thus, two sources generate a combined packet arrival process of 6 packets per second.
- **Number of Servers**: The multiplexor consists of a single outgoing line, thus the number of servers is unity.
- **Buffer Size**: We will assume that the buffer space within the multiplexor is sufficiently large to be considered as infinite. If it is not possible to make such an assertion, then the system is now a lossy system and an alternative model must be used.

Combining all of these assumptions allows us to use the M/M/1 delay equation to consider the dimensioning problem discussed previously. We will consider the problem in three levels of complexity.

Stage 1: In order for the multiplexor not to saturate, the key assumption that the arrival rate must not exceed the departure rate must hold (i.e. $\rho < 1$). Thus, the total offered load must not exceed 20 packets per second. As each source generates 3 packets per second, then we are limited to connecting a maximum of 6 sources to the multiplexor; connecting 7 sources would give an offered load of 21 packets per second and result in saturation. Connecting 6 sources to the multiplexor, would give an estimated average delay of $T = \frac{1}{20-18} = 0.5$. This is quite a high value and may not be acceptable. We will now consider the case when delay is constrained.

Stage 2: A further restriction is now applied to our multiplexor such that the mean delay experienced by a packet should not exceed 250 ms. Using the M/M/1 equation we can work back and determine the maximum offered traffic: $T = \frac{1}{20-\lambda} < 0.25$ s. Thus the maximum number of sources that can be supported cannot now exceed 5 ($5 \times 3 = 15 < 16$). With 5 sources connected, the average delay is estimated as $T = \frac{1}{20-15} = 0.2$ s which is better than the specified limit. The introduction of a simple delay constraint has reduced the number of sources that can be supported by 1. A further constraint could applied by recognising that the delay figure is just a mean; 50% of packets will experience a delay less than a given value while 50% will experience a delay greater than the mean. Service levels are usually expressed in terms of a mean delay and an upper limit that a specified number of packets must not exceed. The third step now considers such a situation.

Stage 3: In the multiplexor, a third criteria is now set such that 90% of all packets must be transmitted within 0.325 seconds. In order to apply this criteria, we need to recognise that delay is itself a random quality and will have its own probability density function. In the case of the M/M/1 system, it is possible to show that indeed delay follows a negative distribution such that $P(x \leq X) = 1 - e^{-mX}$ where $m = \mu - \lambda$. The proof of such a statement is not trivial and is not a matter for this book. However it fits in with what one would intuitively expect given that the occupancy of the queue is essentially described by a geometric series. We have shown that the mean delay associated with an M/M/1 queue is $T = \frac{1}{\mu-\lambda}$, thus we can combine this with the general expression for a negative exponential distribution to consider the third part of the problem.

In the previous part, we suggested that only 5 sources could now be attached to the multiplexor with an associated average delay of 200ms. Furthermore, we can see that probability that a packet's delay achieves the original stated target of being below 250ms is given by $P(x \leq 0.25) = 1 - e^{-0.25/0.2} = 0.713$. Currently, with 5 sources attached ($m = 5$), only 80% of packets meet the newer stricter upper limit. The target is 90%. The new criteria that 90% of packets should have a delay below 350ms can them be expressed as $P(x \leq 0.325) = 0.9 = 1 - e^{-0.325m}$. Solving, gives $m = 7.085 = \mu - \lambda = 20 - \lambda$. Thus $\lambda = 12.91$, which means that only 4 sources should now be connected to the multiplexor.

Buffer Size: We have assumed that the multiplexor buffer is sufficiently large as to be infinite. It is now possible to consider, briefly, the implications of such a statement. In the first stage of the dimensioning problem we indicated that 6 sources could be attached to the multiplexor. In this case the utilisation of the system is $\rho = 0.9$, which using either Little's Law or by direct calculation, $\frac{\rho}{1-\rho}$, gave an average occupancy of 9 packets. Clearly if the buffer has space for 100 packets then our assumption of infinite capacity is reasonable strong. If, however, the buffer only has space for 20 packets then clearly there is a strong chance (> 10%) that saturation will occur and packets will be discarded. In such a case, an alternative, loss-based model should be applied. In the other two scenarios, i.e. 5 and 4 users, the mean system occupancy is much less (3 and 1.5 packets respectively) and our infinite assumption holds.

4.7   Link Properties

As well as the connectivity of the network and the switching capacity of the nodes within it, the properties of the individual links also define the capabilities of the network. Links can be uni-directional or bi-directional. In a bi-directional link, transmission capacity is provided in both directions between the two nodes interconnected by that link. A uni-directional link provides capacity in only a single direction; one can think of a bi-directional link as two unidirectional links each working in opposite directions. In addition, a link may also be described as being a full duplex or half-duplex. A full duplex link is essentially a point-to-point bi-directional link providing equal transmission capacities in both directions simultaneously. A half-duplex link is one that can provide transmission capabilities in either direction but not at the same time, i.e. a bi-directional link whose direction of transmission can be toggled as appropriate.

The major performance parameters of a link are:

- **Throughput**: the rate at which information can be sent over the link.
- **Propagation delay**: the time between when a message is made available at the transmitter for transmission until delivered by the receiver.
- **Corruption**: the proportion of messages lost or corrupted during transit. Since corrupted messages are usually discarded, this property is often termed 'loss'.

There is a trade-off between these quantities, particularly at the physical layer. Raising throughput may increase queuing over part of the link and increase delays. It is also very likely to increase corruption.

4.8   Internetworking

Until now we have assumed that the network carrying information is a single homogeneous entity. This is not really likely. In reality, there are a plethora of network types and protocols in use and which for various reasons require to be interconnected.

Internetworking is achieved by using a device known as a *gateway* (see Figure 4.23) to act as the interface or conduit between the two networks being interconnected. Since different networks will operate in different ways, the gateway will have to address a number of issues, such as the services carried, the protocols in use, quality of service provisioning, packet formats, addressing formats, and so on.

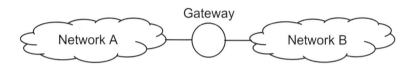

Figure 4.23   Network interworking through a gateway

Internetworking can be viewed from either a network-by-network or an end-to-end transmission service perspective. The network-by-network approach forms an effective virtual circuit between stations by splicing together reliable virtual circuits within each network. The end-to-end approach only assumes that each network supplies a datagram service. A common end-to-end protocol is then used to produce a reliable service if required. An Internet protocol defines a standard format for passing data between networks. An example of such a protocol is Internet Protocol (IP).

Four gateway types can be defined: repeaters, bridges, routers, and higher level gateways. However, in practice many gateways perform multiple functions, with bridge and router functions in particular often being combined.

4.8.1   Repeater

Repeaters provide physical layer connections between segments of the same network (see Figure 4.24). They are dumb (no software) and just copy bits from one segment to another. This means that networks A and B in Figure 4.23 are logically the same network. A hub is a repeater in a physical star/logical bus LAN which broadcasts bits between Network Interface Cards (NICs), which may be either individual stations or other higher level hubs.

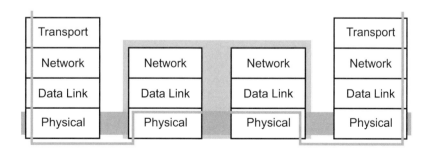

Figure 4.24   Protocol stack for a repeater

4.8.2   Bridge

Bridges connect networks at the data link level, as shown in Figure 4.25. They can be used to connect two identical network types or to interconnect two dissimilar networks which have the same network layer. An example of the latter is the family of IEEE 802 LAN standards which have different media access control protocols but the same network layer. Bridges are smart and can be configured to selectively copy frames from one segment to another. They are often transparent to the network and therefore do not require an address. However, one is often assigned to allow network management functions. MAC addresses are used to send frames between LANs connected by bridges. Multi-port bridges can be used to bridge between different LAN types.

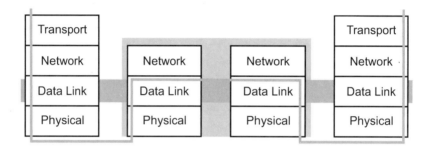

Figure 4.25   Protocol stack for a bridge

4.8.3   Router

Routers are used to interconnect networks that have the same transport layer but have different network layers (see Figure 4.26). A router has an address on each of the interconnected networks. Frames that use a network address (or an Internet address) are routed between networks using a router. Such devices can allow the interconnection of different network types, e.g. an IEEE LAN to an X-25 network.

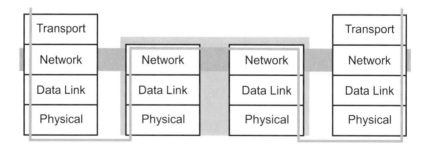

Figure 4.26   Protocol stack for a router

4.8.4 Higher Level Gateway

Higher level gateways provide connectivity between services at either the transport layer or application layer, as shown in Figure 4.27.

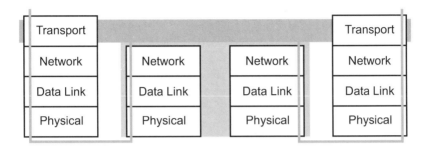

Figure 4.27 Protocol stack for a gateway

4.9 Routing

A key function of the network layer is to route packets between source and destination nodes. Choice of a path through the network might only be made once at the start of a session in the virtual circuit case or on a packet by packet basis for the datagram case. In the latter case, it might only be the source node which makes a routing decision or alternatively each intermediate node might have to route each packet.

The routing algorithm is part of the network layer software which determines the path taken by an outgoing packet. For example, in the relatively simple network shown in Figure 4.28, there are six possible routes connecting X and Y. The path chosen may only represent which outgoing line to use, of which there are two at X, or it could represent a detailed, multi-link path between source and destination. In other cases, the node may take no routing decisions at all and will only ensure that the packet is forwarded to a link. This is the case at node Z.

There are a number of desirable criteria; ideally we would like a system which is correct, simple, stable, robust, fair and optimal. The first two are self-evident – any algorithm used should work correctly and be as simple as possible. Any real network will be subject to failures – link, node, hardware and software – and any routing algorithm should be able to withstand changes in the network state without having to abort or restart. Stability is important, as the traffic presented to a network is likely to vary considerably, and these variations should not cause undue variation in the utilisation of links/nodes to the detriment of others. A routing algorithm should endeavour to provide a fair service to all users – individuals should not unintentionally be denied service due to the demands placed on the network by others. This requirement has to be balanced against the desire to optimise the performance of the network. A routing algorithm could, perhaps, try to minimise the delay experienced by a packet while also maximising the utilisation of the network links. These two goals represent a conflict – utilisation can be improved by having large, full queues which in turn implies increasing delays – so some sort of trade-off will be required.

Adaptive algorithms base their routing decisions upon the current state of the system. Non-adaptive algorithms do not take into account the system state when selecting a suitable path.

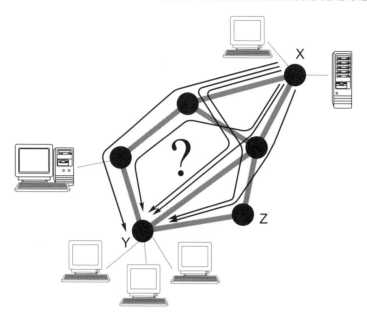

Figure 4.28 Routing

It would be expected that if an adaptive algorithm reacted well to traffic changes or the network topology, it would out-perform a non-adaptive strategy. That said, the realisation and implementation of adaptive algorithms are considerably more complex.

In a centralised routing, a central node/processor will collate all relevant information and compute all the relevant routes – optimal and/or alternative – and pass this information to each node within the network. De-centralised methods can be sub-divided into two types: isolated and distributed. In the former, a node uses only the information that it has itself gleaned about the state of the network. Distributed routing refers to adjacent nodes exchanging information to update routing tables. Isolated schemes are relatively simple to implement but do not, generally, yield optimum routes. Centralised routing is difficult to maintain if the network changes. It is possible to combine the two by using dynamic de-centralised methods and recording a centralised 'router of last resort' should a packet to an unrecognised destination be encountered. This increases the chances that a route will always be found. It is also possible to look at routing in terms of either obtaining the single best possible path for data through the network or to recognise that there can often be more than one way for the data to travel to its destination. Multipath routing has the advantage that it is inherently more reliable than single (shortest) path routing but does not always use the best possible routes.

4.9.1 Shortest Path

The shortest path between two nodes represents the best or optimal path which can be used to transfer information between each party. What the shortest path actually is depends upon the criteria used to judge this. Many different metrics can be used: number of links, distance, delay, bit-rates and cost. The shortest path between two nodes using the number of links as the only criterion may not necessarily be the same if delay was the criterion.

In determining the best/shortest/optimal paths in a network it is necessary to have some idea of the relative merits of each link as they are unlikely to all be equal. This is achieved by labelling each link with a weighting which has to be computed using the metrics of interest. The particular metric can, of course, vary from network to network. Links with lower weightings are more suitable than ones with a higher weighting. Note that if the number of links is the only metric, then all links are equal and would have the same weighting.

The next step is, given the link weighting, to obtain the optimal or shortest path. In simple networks this can often be found by inspection. Complex networks present problems and inspection is not really a suitable technique to use as a base to control such a complex and costly resource. Formal methods should be used and there are a number of algorithms developed to perform such a task.

Algorithms such as Bellman-Ford and Dijkstra take a single node and calculate the shortest paths between this and all other nodes in the network. Other algorithms, such as Floyd's, find the shortest paths between all pairs of nodes within the network.

4.9.1.1   Dijkstra's Algorithm

Dijkstra's Algorithm, which finds the shortest path from a given node to all other nodes in a network, proceeds as follows.

1. Label all the nodes except the starting node with a distance, initially ∞, and an approach node, initially $-$.
2. Beginning from the chosen starting node, choose the node with the lowest cumulative weighting; this node if 'fixed'.
3. Label its adjacent nodes with the node name as the approach node and the cumulative distance from the starting node.
4. If a node is already labelled, its label is only replaced if the new cumulative distance is less than the existing cumulative distance.
5. Continue until all nodes are fixed.

After all nodes are fixed, the resulting cumulative weights are the shortest paths from the source node. The path itself can be calculated by reading back the approach nodes at each node. The following example shows the algorithm in operation.

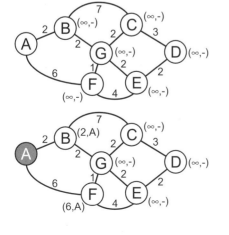

Each node is initially labelled $(\infty, -)$

Starting with node A, and fixing it, relabel its adjacent nodes, B and F, with their distances

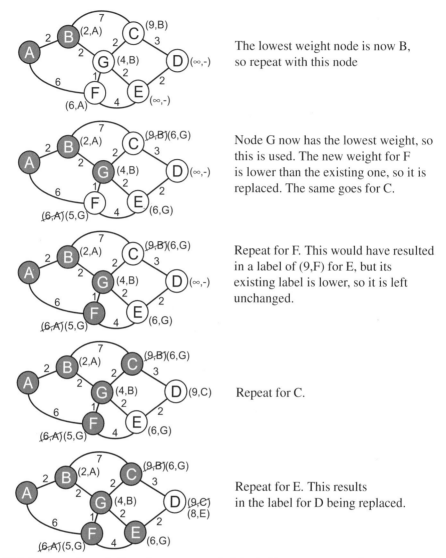

The lowest weight node is now B, so repeat with this node

Node G now has the lowest weight, so this is used. The new weight for F is lower than the existing one, so it is replaced. The same goes for C.

Repeat for F. This would have resulted in a label of (9,F) for E, but its existing label is lower, so it is left unchanged.

Repeat for C.

Repeat for E. This results in the label for D being replaced.

This leaves only node D to fix, so we are finished. The shortest path from A to D therefore has cumulative weight 8, and reading back the labels starting from node D, leaves a route D←E←G←B←A.

4.9.2   Flooding

This is perhaps the crudest of routing techniques, where any packet received by a node is immediately forwarded via all outgoing links (except the source link). This always finds the best route, as since all routes are attempted, the information will arrive at the destination first by the shortest route. However, an obvious and significant disadvantage is that this will result in a very large number of packets circulating in the network, so some mechanism is required to dampen the process.

One method would be to have a link count field in each packet which would be decremented by each receiving node. If the field gets to zero, then the packet is discarded. The value of this count field should lie between the number of links in the optimum path between source and destination, and the diameter of the network, so preventing packets from circulating endlessly within the network.

A second technique could be to identify each packet uniquely and have each node keep a record of all the packets they have processed. If a node then receives a packet it has previously forwarded then the duplicate can be discarded.

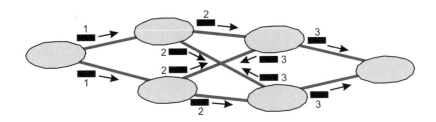

Figure 4.29   Routing with the flooding approach

Variations include selective flooding where packets are forwarded only over links which are in the 'right direction', and the Random Walk algorithm where a packet is forwarded over a single link chosen at random – the packet will eventually get to its destination.

4.9.3   Distributed Routing

Distance Vector Routing and Link State routing are both examples of dynamic distributed routing where routing nodes obtain information about the state of the network or estimates for routes from other nodes in the network and then use this to derive a routing table or vector. In a distributed system, nodes interrogate adjacent nodes for information on which to base their routing decisions. Such systems can operate in a direct fashion, asking for specific information at regular intervals.

4.9.3.1   Distance Vector

Distance vector routing is used in the Routing Information Protocol (RIP) which is used for Internet routing.

Consider node A in the network in Figure 4.30. Periodically, node A would ask nodes B and C for information about the shortest routes from these nodes to all other nodes in the network. Node A would also obtain the 'distance' between itself and the adjacent nodes and would use the resulting information to compute the shortest paths between itself and all other nodes. Some of the required information can be obtained indirectly. If delay is the metric upon which routing decisions are made, then node A can find the distance to the adjacent nodes by sending 'echo' packets to each node. Once these packets are forwarded a timer is started. This timer is stopped when the packet is echoed back and received correctly by node A, thus allowing node A to compute the delay associated with that link.

In the example network shown in Figure 4.30 it takes 10 and 12 time units for nodes B and C (respectively) to respond and send their current distance vectors to Node A. Node A can now infer that it takes 5 and 6 time units for a packet to travel to nodes B and C respectively. The estimates of 6 and 4 in B and C's tables are ignored as the estimates of 5 and 6 are considered to be more current. The derivation of a new table for node A is now relatively simple. Consider the case to determine the route from A to D, node A estimates that if it forwards packet destined for D directly to node B then it will take 5(AB) plus 4 (A – D) = 9 units of time. The estimate going via node C is 6(AC) plus 5(C – D) = 11 time units, thus the first choice via B is adopted. The procedure is repeated for all other non-adjacent nodes within the network. Traffic between AE is forwarded via node B with a total distance of 11 (5 + 6) rather than via node C with a total distance of 12 (6 + 6). Finally, packets destined for node F are forwarded not via node B, a route which gives a total distance of 18, but via the shorter path offered via node C of 16 time units.

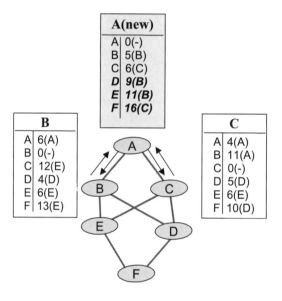

Figure 4.30   Distance vector routing

The main disadvantage with this scheme is that it reacts slowly to bad news such as node outages, and continues to use inappropriate routes as a result. In larger networks, changes in route lengths can take a significant time to permeate to all other nodes – if at all! In particular, distance vector routing may continue to send packets along towards links that no longer exist and such packets thus circulate around the network. On the other hand, the scheme responds well to good news, i.e., improvements in routes.

4.9.3.2   Link State

Link State routing is an improved alternative to distance vector routing and is now widely used in practical routing devices. It is used in the OSPF (Open Shortest Path First) protocol, which is increasingly used for Internet routing and addresses some of the shortcomings of RIP. It overcomes the disadvantages of the previous scheme in terms of the 'count to infinity' issue and solution convergence.

In broad terms, this is a simple autonomous scheme with four steps:

- **Identify and contact adjacent nodes**: It is necessary for each node to know who it is connected to at any given time. Such information is obtained by the transmission of special packets. The key issue is that each node must have a unique identity.
- Determine cost to each adjacent node: Each node then attempts to determine the distance/cost associated with each of its links and then creates a table with this information
- **This table is then to be forwarded to all other nodes**: In this latter stage, it is important that each node receives these packets or starts to use these at the same time as all other nodes, otherwise routing inconsistencies may occur.
- **Each node then computes best paths to all other nodes**: Formal algorithms such as Dijkstra's can used by each node to determine the shortest path from it to all others.

Issues with this scheme include ensuring that link state tables are created/updated at the appropriate times, the updated packets are sent/received correctly and what to do when nodes are lost. Figure 4.31 shows an illustration of link state routing where the highlighted entries indicate the route a packet would take between nodes A and F and F and A respectively.

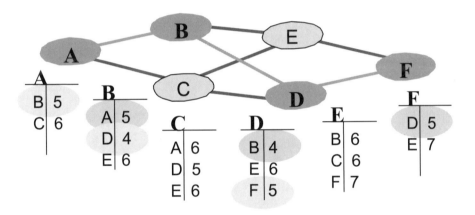

Figure 4.31   Link state routing

4.9.4   Broadcast Routing

The final form of routing covered here is broadcast routing. This is where a node specifically wishes to send the same data to all other nodes. Distinct packets can be sent to every destination node with each packet being uniquely identified by its address field. This can be wasteful of system capacity and time-consuming, requiring each node to have a complete list of all nodes within the network.

Flooding can be used, although (as in the point-to-point case) it generates too many packets and requires significant bandwidth.

Broadcast packets can contain more than one destination within the address field. When a packet arrives at a node, the node checks the routing table and destinations in order to determine the output lines needed. Copies of the packet are then made with the address field containing only the addresses of the destinations which are relevant to the output line on which the packet is forwarded. On receipt of such a packet, a destination node will remove its address from the address field and repeat the copying and forwarding procedures. Eventually a broadcast packet will only contain one address – the final node – and thus broadcasting will have been completed.

If each node has knowledge of the broadcast packet source's optimal paths, i.e. sink tree, then the packet can be forwarded only to adjacent nodes if they form part of an optimal path. Such a scheme represents the most efficient way of transmitting broadcast packets but, given the dynamic nature of packet networks and their sheer size, such information may not be practical to obtain.

Alternatively, a node will forward a received broadcast packet only if it arrived on the link that is normally used by that node to transmit to the source of the broadcast packet. Otherwise the packet is assumed to be a duplicate and is discarded. This is simpler than the previous algorithm as each node does not need to know anything about the state of the system, i.e. sink trees. Such a technique is called *reverse path forwarding*. Figure 4.32 shows a simple example of how Reverse Path Forwarding operates and compares it with a broadcast scheme based upon a sink tree.

The comparison here is not so obvious as the network is not dense so it is difficult to see the merits of such a scheme. However, if the network is larger then one can see that reverse path forwarding will not yield a solution as good as a sink tree scheme but will be much better than straightforward flooding with a hop limit. If the latter scheme was applied to the network shown (assume a hop count limit of 3) in Figure 4.32 then 10 packets would be sent compared to 8 with reverse path forwarding.

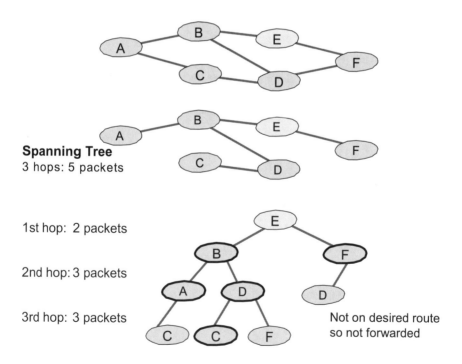

Figure 4.32   Reserve path forwarding

4.10   Congestion

Until now, we have assumed ideal behaviour on the part of nodes. Packets arriving at a node are queued until they can be sent on. If packets arrive at the node at a rate which approaches the node's capacity, packets will build up in the buffer and the mean delay will increase towards infinity. In practice, the buffer would have a finite length and packets would be lost or re-routed. This congestion in the system can result from routing decisions, or simply an overwhelming traffic load. Congestion tends to be self-perpetuating; as one session loses packets it will, after a suitable interval, attempt retransmission, which further increases the number of packets in the system, further increasing the congestion. This means that throughput in a congested system may actually reduce as congestion increases (see Figure 4.33).

A clear distinction should be made between routing control and congestion control. A routing algorithm can reduce the likelihood of congestion, but should it occur, then a congestion mechanism will be required to alleviate the effects. The routing algorithms are likely to aggravate the situation unless there is a congestion mechanism included.

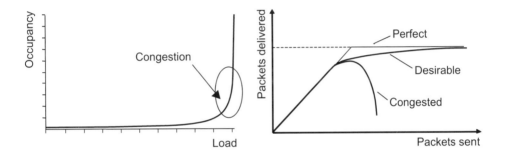

Figure 4.33   Effective of congestion

4.10.1   Admission Control

Virtual circuit calls reserve space in advance in each buffer (in their path) so that there will always be a place to store incoming packets in each node. If a virtual circuit cannot be allocated space in the buffers, then the call is rejected. This ensures that calls currently in progress remain relatively unaffected by any increased offered load.

4.10.2   Load Shedding

This is the reverse of the previous technique, such that no space is allocated in advance and all calls compete for access to the buffers. As a buffer approaches its capacity, newly arriving packets are discarded. Other layers/functions will in turn deal with the effects, i.e. organise retransmission attempts. Clearly, there are limits to the discarding of packets, and rules have to be established to determine whether a packet is discarded or kept. There is no advantage to be gained in discarding acknowledgement packets, and nodes usually ensure that there is always space to store one or more acknowledgements. Discarding does not necessarily have to commence only when the buffers are completely full; load shedding algorithms often start to selectively discard packets at around 80 or 90% of capacity.

4.10.3   Flow Control

This is a feedback mechanism whereby special packets are generated by a node whenever the load on that node reaches certain levels. These choke packets instruct the packet sources to reduce transmission rates by some value, thus preventing the node from becoming congested. Such a method has the advantage that it only becomes operative under high loads and does not affect network performance under normal conditions. The load on the node can be monitored by considering its instantaneous utilisation and estimating the average utilisation or considering the queue occupancy. The following mechanism can be used to monitor the utilisation of a link;

$$U_n = aU_{n-1} + (1 - a)f$$

where f represents the instantaneous line utilisation (i.e. $f = 1$ if the line is busy or 0 if it is not), and a represents a constant which determines how fast the node forgets previous history. U_n represents the new estimate of average utilisation whilst U_{n-1} is the previous one. The effectiveness of this mechanism depends upon the estimates of average utilisation and how quickly it can react to changes in system state.

   The system can be applied to both datagrams and VC, and on an end-to-end basis as well as link by link. It is therefore possible to have a mechanism that would monitor a variety of system parameters over one or more nodes or links, thus identifying potentially congested paths as well as links.

4.11   Error Control

Error correcting coding is included in a transmission system to protect against errors introduced by the transmission medium. Three strategies can be identified for error control:

- **Error concealment**: Here errors are detected and the corrupted information identified so that it can be discarded. The remaining information is available and in some cases can be used to mask the missing corrupt data. This can be done by repeating a previous sample, muting the corrupt sample, or by trying to interpolate from the surrounding values. Error concealment works well in speech and audio systems as long as the error rate is low. It can be also used for images in some cases, but not for general data.
- **Automatic repeat request (ARQ)**: As before, errors are detected, but in this case the transmitter is informed and requested to send the data again. ARQ has a number of disadvantages; it requires a feedback channel to the transmitter, the transmitter has to store the data until it has been informed that it has been received correctly, and, in the case of an error a delay is introduced to the transmission of the data. However, ARQ schemes can achieve very low error rates with relatively low overheads and complexity.
- **Forward error correction (FEC)**: In this case additional message symbols are added to the message being sent so that it is possible for the receiver to reconstruct the message in the event that part of it was corrupted. FEC is computationally complex and requires an additional transmission overhead but has the significant advantage of not requiring a feedback path to the transmitter. Also, little additional delay in introduced over that involved in transmitting the additional symbols.

It is possible to use combinations of these strategies, with FEC being attempted first by the layer layers, and ARQ or error concealment being employed if this fails. Error concealment can only be performed by the application because it knows the context of the data. The strategy most appropriate for the transport layer is ARQ.

4.11.1   Error Detection

In order for the system to respond to errors, it must first be aware that they have occurred. There are two possible problems. The first is that a frame which has arrived at the receiver has be corrupted by one or more errors. The second possibility is that the frame may not arrive at all. The second problem can be guarded against by including sequence numbers in each frame. If a frame with a particular sequence number does not arrive, it can be considered lost.

Detecting errors is more complex. A function is calculated over the message by the transmitter and added to the frame as a Frame Check Sequence (FCS). When the receiver gets the frame, it calculates the same function, and compares that to the FCS. If the calculated value is different from the one that was sent, it can be concluded that some alteration has been made to the message between the time the two functions were calculated, i.e., between the transmitter and the receiver, so an error has occurred.

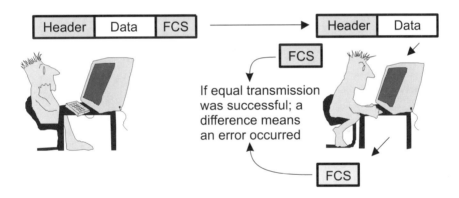

Figure 4.34   Frame check sequence

The danger is that an error could occur in such a way that the function could still produce the correct result. For example, if the function was a simple addition of the 16 bit words in the frame, as in a UDP packet, if one error added to a word and another subtracted from another word by the same amount, the error would not be detected. This possibility can be reduced and even eliminated for a given number of errors by careful design of the function based on the principles discussed in Section 5.4. Cyclic redundancy checks (CRC) (see Section 5.4.3.2) are often used. The 16 bit CRC defined by the CCITT which is commonly used can detect:

- all error bursts of length up to 16 bits, where a burst is defined as a block starting and ending with an error, whose intermediate bits may or may not be in error
- 99.997% of bursts of length 17 bits
- 99.998% of bursts longer than this
- all possible combinations of 3 or fewer random errors
- all possible combinations of odd numbers of errors

4.11.2 ARQ

There are three types of ARQ – stop and wait, go back N, and selective repeat. The 'stop and wait' system is the simplest form. Each message is transmitted from the source to the destination, and the destination acknowledges the message back to the source (see Figure 4.35). The source does not transmit another message until it receives the acknowledgement for the previous message. During this time the transmitter is idle, leading to the term 'idle ARQ'. To guard against lost acknowledgements, a timeout is used, so if the source hears nothing from the destination, it transmits the message again.

Channel *utilisation* is the proportion of time the channel is active. If the frame consists of a header of h bits and a message of m bits, acknowledgements contain a bits, the channel transmission rate is B bit/s, and the propagation delay, the time it takes for messages to travel from the transmitter to the receiver, as τ, in an ideal error-free link it will take $(h+m)/B+\tau$ seconds for a frame to arrive at the receiver. The receiver will then acknowledge this frame by sending an acknowledgement which will be received by the frame source in $\frac{a}{B}+\tau$ seconds. The node can then transmit the next frame. Thus, the total time required to transmit a single frame is $\frac{h+m+a}{B}+2\tau$ (see Figure 4.35). Over this period m message bits are transmitted, so the useful time is $\frac{m}{B}$. The utilisation is the useful time/total time, is therefore $\frac{m}{h+m+a+2B\tau}$. High utilisation therefore depends on keeping the header and acknowledgement short, and the propagation time low.

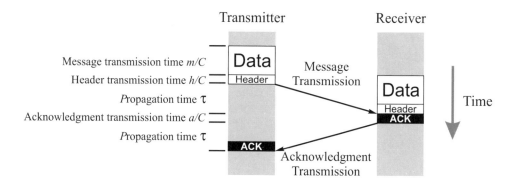

Figure 4.35   Stop and wait ARQ

To overcome the inefficiency associated with stop and wait ARQ, it is possible to transmit new frames up to a given limit without waiting for any received ACKs. The number of outstanding unacknowledged frames is known as the window and is fixed to some maximum value, and leads to the idea of a sliding window protocol. The stop and wait ARQ protocol is often referred to as a sliding window protocol with a window size of 1. Use of windows in this manner provides a degree of flow control and prevents a transmitting node overwhelming a receiver (quite likely in case of retransmissions), as well as ensuring an upper limit on the buffer space required.

To deal with lost acknowledgements, a timeout can be used for each frame as in stop and wait ARQ. If an acknowledgement has not been received by the time the timer times out, the transmitter assumes that the frame is lost and that another has to be retransmitted. Alternatively, breaks in sequences can be used to indicate that a frame has been lost.

There are then two retransmission options available; either the transmitter transmits only the frame that was in error (selective retransmission) or it also retransmits all the frames that had been transmitted after the lost frame (go back N retransmission).

In a selective scheme, the transmitter resends only the frame that was in error. This has the advantage that it is less wasteful of link capacity but relies upon the receiver being able to ensure that frames are delivered to the network layer in the correct order. This means that it has to buffer correctly received frames while it waits for retransmitted frames.

In a go back N scheme, the transmitter 'goes back' to the lost frame and re-sends all the frames from that point on. The advantage of this approach is that no re-sequencing or buffering is required at the receiver, but since it requires the retransmission of frames which may have been received correctly, it is not quite as efficient as selective retransmission in the case of errors.

If the propagation delay is small and therefore few frames are transmitted before an acknowledgement is received, the number of frames affected will be small, but over long distances with networks with large numbers of links and a significant propagation delay could occur, and the order of re-transmission would be preserved at the cost of lost capacity.

In a go back N system the receiver uses sequence numbers to identify when an error has occurred or when it has received duplicate frames. If a frame has been received correctly it will have a sequence number N, and the next frame received should have sequence number $N + 1$. If the number received/detected is $N + 2$ then a frame has been lost.

In sliding window protocols, the role of sequence numbers is critical. As a frame is transmitted it is given a number which identifies when it was output in relation to other frames. These numbers are used by the receiver to ensure that the frames are passed onto the network layer in the correct order.

The range of sequence numbers allowed is a function of both window size and retransmission method. In the case of a sliding window protocol with a window size of 1, only a 1 bit sequence number is required. In general with a window of K frames, the range of sequencing numbers required is 0 to $2K + 1$ for a system utilising selective retransmission and 0 to $K + 1$ for 'go back N' systems. In practice, sequence number fields within frames are preset, usually to 3 bits or 8 bits.

Calculating the utilisation of sliding window protocols is more complex. Two cases can be identified. The first is the large window case where the window size is sufficient to allow continuous transmission under error-free conditions. In the small window case, the window size is not large enough to allow continuous transmission, so there is effectively a stop and wait ARQ scheme only with the transmission of more than one frame at a time.

It takes $\frac{h+m}{B}$ seconds to transmit a frame. It takes $2\tau + \frac{a}{B}$ to receive an acknowledgement. This means that $\frac{(2\tau+a/B)}{((h+m)/B)} = \frac{2B\tau+a}{h+m}$ frames can be transmitted while awaiting an acknowledgement. If the window size is greater than or equal to 1 more than this value, then continuous transmission is possible, and the utilisation is $m/(h+m)$. The additional 1 is due to the original frame for which the acknowledgement was for. In the small window case, with a window size, W, $W < (2B\tau+a)/(h+m)$. For each interval of time $\frac{h+m+a}{B} + 2\tau$ seconds, W frames can be sent, which increases the error-free efficiency by a factor of $\frac{W(h+m)}{(h+m+a+2B\tau)}$. This will in turn result in a small window link efficiency of $\frac{m}{h+m}\frac{W(h+m)}{(h+m+a+2B\tau)}$, W times the simple stop and wait ARQ case.

Figure 4.36 illustrates the different types. In each case, the channel suffers from errors at the same point. For 'stop and wait', the 1st, 2nd and 3rd blocks are transmitted and received correctly (the first channel error burst occurs during a wait period). The 4th message is corrupted by errors, and has a negative acknowledgement. It is therefore retransmitted. In the 'go back n' case, the first error burst corrupts the 2nd message block, but the source is in blissful ignorance until it receives an acknowledgement, by which time it has transmitted the 3rd and 4th message blocks. The source has to go back to the 2nd block and transmit it again. In this case, n must be 3 or more. Had it been only 2, the source would have had to stop after transmitting message 3 until it received the acknowledgement of block 2. The selective repeat system also assumes a minimum buffer of at least 3, and only retransmits errored blocks. Note that this system transmits 11 blocks in the time the other systems managed only 7 or 5.

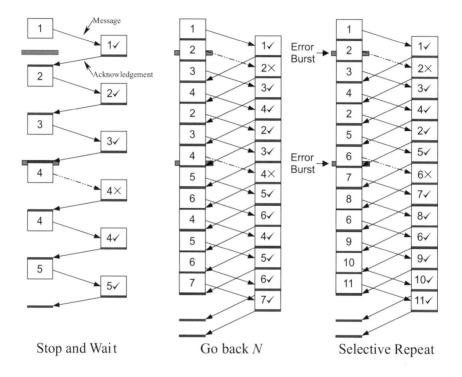

Figure 4.36 Operation of the different types of ARQ

If a block is in error, it must be retransmitted. It is therefore sensible to keep the blocks short, to reduce the amount of information retransmitted. However, short blocks mean more messages, more acknowledgements, and since the header has a minimum size, reduced utilisation in the error-free case.

The utilisation of stop and wait ARQ when errors occur can be calculated as follows. There will be a timeout which will trigger a retransmission if an acknowledgement is not received. The minimum value for this time is $2t + a/B$, the amount of time it would take for an acknowledgement to return normally. In this case, frames will be transmitted at the same interval whether or not an error occurs. Let P_f = probability of a lost frame and P_a = probability of lost ACK. The probability that a frame is correctly received, $P_s = (1-P_f)(1-P_a)$. The useful throughput per frame is therefore $m \times P_s$. Utilisation is therefore $mP_s/(h + m + a + 2B\tau)$.

A similar approach can be used for the selective repeat ARQ, if we allow the simplification that retransmissions of lost packets are not themselves lost to the point that the system has to reduce the number of packets it sends due to full buffers. This equates to the condition that the error rate is low. In this case, utilisation is P_s of the error-free case, and so is $mP_s/(h + m)$ for the large window case, and $mWP_s/(h + m + a + 2B\tau)$ for the small window case.

Unfortunately, we cannot apply this approach directly for go back N ARQ because in addition to the fact that more than one frame will have to be resent (N frames in the large window case, W in the small window case), failure of the retransmission of a failed frame will cause the protocol to stop. However, if we allow the simplification that this does not occur, then each lost frame results in the retransmission of N or W frames. The average number of transmissions required to see the frame transmitted successfully be $1/P_s$. The average number of retransmissions is one less than this, i.e., $1/P_s - 1$ so the total number of frames transmitted will be $(1/P_s - 1)N + 1$ for the large window case, or $(1/P_s - 1)W + 1$ for the small window case. The utilisation will be reduced by this factor, giving $\frac{m}{h+m}((1/P_s - 1)N + 1)$ for the large window case, and $\frac{mWP_s}{h+m+a+2B\tau}((1/P_s - 1)W + 1)$.

Figure 4.37 shows the relative performance of stop and wait ARQ and selective repeat ARQ for a system with a transmission rate of 1Mbit/s, 40 bit headers and 40 bit acknowledgement packets. Two cases are shown: a propagation delay of $10\mu s$ as would be likely in a local system and a propagation delay of 10ms representing a fast multi-hop distance link. Country-wide links can have propagation delays in the order of a hundred milliseconds, rising to many times that on continental links if satellites are used (see Section 6.3.3.4). The probability that a bit is in error on either the transmitted data or its acknowledgement is 10^{-4}.

It can be seen that in all cases utilisation improves towards the maximum and then falls away quickly. The initial increase is because increasing the packet size reduces the signalling overhead proportionally. However, because the probability of a bit error is fixed, larger packet sizes are less likely to be received correctly, and towards the right-hand side of Figure 4.37 these factors begin to dominate. With a bit error rate of 10^{-4} for example, only just over a third of packets will be received correctly when the packet size is 10000 bits. Selective repeat ARQ performs best, and this is always the case, although for large packet sizes, and therefore large error rates, large buffer sizes are required to maintain the performance shown, and performance will be reduced towards the stop and wait case if the buffer size is limited.

The performance of go back N ARQ tends towards that of selective repeat ARQ for low error rates and that of stop and wait ARQ for high error rates. Since in practice systems are designed to have a low probability of retransmission, this means that go back N ARQ is often a good choice. Note that the approximation of performance of the go back N case given above does not hold in this example above a packet size of about 1000 bits, because the probability of retransmission exceeds 10% and the probability of having to retransmit retransmissions stops being insignificant. Beware – applying the formula to high error rates gives the false impression that go back N ARQ performs worse than stop and wait ARQ, which is clearly impossible.

Selective repeat ARQ is not significantly affected by propagation delays (assuming a large enough buffer) but the same cannot be said for stop and wait ARQ. With short propagation delays, stop and wait ARQ has reasonable performance, but if the system has to wait relatively long for an acknowledgement, the time spent idle becomes very significant. Selective repeat ARQ outperforms stop and wait ARQ even for low propagation delays, but the difference is not very large at low error rates. For higher bit error rates the difference increases, so

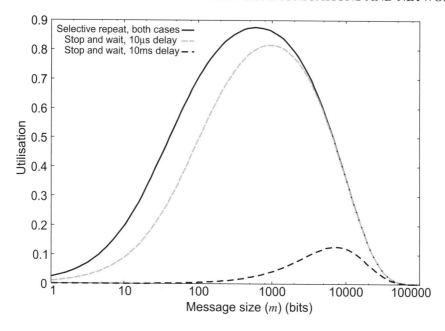

Figure 4.37   Relative performance of selective repeat and stop and wait ARQ Schemes

for example with a 10^{-3} bit error rate, the maximum utilisation of selective repeat ARQ is about 65% whereas for stop and wait it is about 50%. However, utilisation in either case is not particularly high, and a better strategy would be to use some forward error correction to reduce the high error rate in the lower layers before the user of ARQ. A discussion of the relative merits of the two approaches can be found in Section 5.5.

4.12   Transport Layer Services

The network perspective of the communication system involves two layers of the OSI stack, the transport layer and the network layer. The network layer undertakes routing, and the routing mechanism must obviously consider the underlying network or networks. The transport layer, on the other hand, provides a transport protocol to transport messages from one process on the network to another and is completely independent of network type. Essentially, the job of the transport layer is to provide end-to-end connectivity that is independent of network infrastructure. It has to deal with addressing, connection set-up, flow control and crash recovery.

To cater for different network types (e.g. connection oriented and connectionless) and to serve different application demands, the transport layer offers a number of different classes of service. Both the ISO and Internet models define transport service classes. The former were designed for telecommunications services rather than packet data services alone, but the rise of the Internet means that although it only provides two classes of services, the connectionless user datagram protocol (UDP) and the connection-based transmission control protocol (TCP), it is this model which is now prominent.

A similar diversity of view can be seen in relation to the services provided by the network layer to the transport layer. In the OSI approach, the network layer can provide 3 different services for different traffic types. In the Internet approach, there is only one, IP. IP is a very simple service and was designed to work over a number of different networks. The approach is basically that of taking the lowest common denominator. However, this allows the underlying network to be simpler, and the popularity of IP means that protocols are being developed for a number of services never originally envisaged as running over IP, such as voice.

Both models support connectionless and connection-orientated services but in a different manner. OSI betrays its telecommunications origins with support for both in the network layer but offering only connection-oriented services in the transport layer. The Internet approach has a connectionless network layer but both connectionless and connection-orientated services in the transport layer.

4.12.1 Addressing

In order to communicate, entities must be able to locate and address their peers. A distinction can be drawn between transport layer addresses and network layer addresses. Transport layer addresses should map onto processes used to identify the actual communicating entity, which is likely to be a particular process on a host. The network layer sends information to a specific terminal, and so network layer addresses refer to the terminal itself.

4.12.1.1 Hierarchical Addressing

Hierarchical addressing uses a series of fields, each specifying location of addressed object more precisely. An example is a telephone number: *<country, city, exchange, number>*. Hierarchical addressing is often used in network layer where it simplifies routing, since the location is implicit in address. However, since the transport layer does not perform routing, this advantage does not apply. In fact, it can be a disadvantage, since the communicating entity moves to new location it cannot take the address with it since it contains a network component. Returning to the example of the telephone network, if a business moves to a new area, it cannot take its telephone number if the number is a standard exchange-based one. New 'personal' numbers, starting '07' in the UK, can be moved as they use the flat addressing structure described below. However, since the telephone network uses hierarchical numbering, the flat address has to be translated into a hierarchical address as the call is set up. This places an additional obligation on the network.

4.12.1.2 Flat Addressing

Flat addresses use a single field with no particular relationship to geography or other hierarchy. This field has to be long enough to provide addresses for all transport service access points on network. Flat addressing requires central record of all allocated addresses to ensure every address ID is unique, and although this record can be distributed, connection set-up is more complicated. Flat addressing allows communicating parties to be more mobile, but routing is more difficult.

4.12.1.3 Initial Connection Protocol

If a system offers many services it is not feasible to permanently assign transport addresses to each. One solution is a procedure called initial connection protocol (ICP) using a process server. Potential users of any service first connect to a process server, which then generates a new process and assigns a transport address for this new process. The process server then disconnects from user and the user connects to new process.

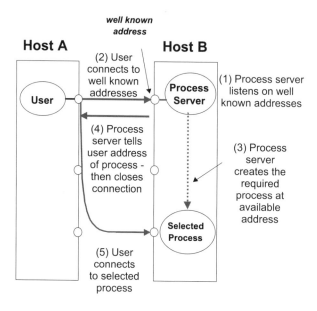

Figure 4.38   Transport address assignment

4.12.2   OSI Transport Service Classes

In the OSI model there are five transport classes of service. The network layer may not be reliable, and there may be *residual errors*, i.e., errors which are still present after being the lower layers have done their processing. In addition, there may be *signalled failures*, in effect an error detected by the network layer which it cannot cope with. The transport layer has to detect residual errors on its own. The OSI model defines three types of network:

- **Type A**: Network connections with acceptable residual error rate and acceptable rate of signalled failures.
- **Type B**: Network connections with acceptable residual error rate and unacceptable rate of signalled failures.
- **Type C**: Network connections with residual error rate not acceptable to the transport service user.

Five transport layer services are defined over these networks as follows:

- **Class 0** is the simple class and is used with type A networks, i.e., networks which provide acceptable performance on their own. It was developed by what was then the CCITT for use with telex. It provides a connection based on network level flow control and with connection release.

- **Class 1** is the basic error recovery class and is used with type B networks. It was developed by CCITT and designed to run on X25 networks where it provides minimal error recovery to network signalled errors.
- **Class 2** is the multiplexing class and is used with type A networks. It is an enhanced version of class 0, and like that service assumes a reliable network service. It supports the multiplexing of multiple transport connections on to one network connection along with individual flow control for each transport connection.
- **Class 3** is the error recovery and multiplexing class and is used with type B networks. It is a combination of classes 1 and 2 providing multiplexing and flow control (class 2) along with resynchronisation (class 1) for failure prone network.
- **Class 4** is the error detection and recovery class and is used with type C networks. Unlike the other classes which all require connection-based network services, Class 4 services can be supported on a connectionless network service. It assumes the network service is unreliable with an unacceptable residual error control, and has flow control, connection set-up and termination, and crash recovery.

The ISO Class 4 transport service and the Internet TCP transport service are similar. However, all ISO transport services are connection-orientated, and there is no equivalent to the Internet UDP transport service.

4.12.3   Internet Transport Service Classes

The Internet model takes a different approach. It basically assumes that the network can only provide an unreliable connectionless service, and only provides two transport classes, TCP, which equates to the ISO Class 4 service, and UDP, which is connectionless. Both these protocols operate over Internet Protocol, IP, a single network layer service designed to provide a level playing field on whatever network is available.

4.12.3.1   User Datagram Protocol

The User Datagram Protocol (UDP) provides a simple connectionless mechanism for applications to exchange messages. While the fact that no connection is established means that the protocol has very low signalling overheads, it also means that there is no error or flow control. For some real-time services with very low delay requirements like voice transmission, a lack of flow control is an advantage, since any lost data would not be repeated anyway. UDP is also used for broadcast messages since a connection-orientated approach is not then appropriate, and for periodic messages like routing table updates where if the data is lost, it does not matter since the existing data can be retained until the next update. Some services, like DNS, which could use TCP usually use UDP for efficiency. Rather than waste time setting up a connection, as well as adding to the load of the host, a connectionless UDP request is made. If the request or its response is lost, another DNS server will be tried after a timeout.

The UDP PDU has four 16 bit fields, shown in Figure 4.39, with the source and destination ports referring to application processes on local and remote hosts. The source port is optional; it is set to zero if it is not used. The length field refers to the total number of octets in datagram including the header.

Note that the UDP segment does not include the address of the recipient, only the port number. This is because UDP is designed for transport over IP, and the IP header, covered in Section 4.13, holds that information. There is still the problem that since the UDP header does not contain that information directly, a UDP datagram could be delivered to the wrong host and the transport layer would be unaware of this fact. To avoid this, the source address, destination address, protocol and length of the IP packet header (see Figure 4.42) are considered to form a psuedo-header which is added to the UDP datagram for the purposes of calculating the frame check sequence. This is the ones complement[1] of the arithmetic sum of the datagram taken 16 bits at a time. If the datagram is delivered to the wrong host, the checksum will fail. The checksum is optional and is set to zero if not required, but if it is used it can be checked by intermediate routers which can drop corrupted packets to save on network load.

Note that this means that UDP is dependent on IP and cannot be used as a transport protocol for other network protocols. That dependency works only one way, however. IP is not restricted to carrying UDP.

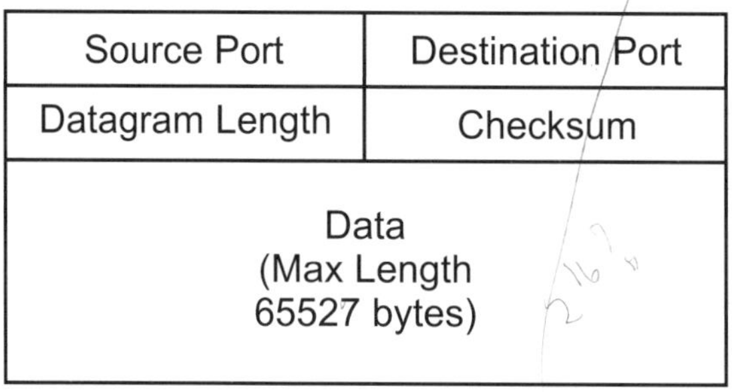

Source Port	Destination Port
Datagram Length	Checksum
Data (Max Length 65527 bytes)	

Figure 4.39   UDP packet

4.12.3.2   Transmission Control Protocol (TCP)

Transmission Control Protocol (TCP) provides a connection-orientated communications protocol designed to work over IP (see Figure 4.40. Like UDP, TCP allows communication between specific processes on each host, so there can be many different connections between two hosts simultaneously. However, since the identification of a connection is done on the basis of ports on each host, there can only be one connection between a given source port and a given destination port on a pair of hosts.

Since IP only provides an unreliable transfer mechanism, and TCP is connection-orientated, TCP must provide mechanisms to ensure that any lost or corrupted data is replaced before delivery to the upper layers. This is done using ARQ mechanisms. TCP has three phases of operation: connection establishment, data transfer and connection termination. A three-way handshake is used in the connection phase because of the unreliable network service.

TCP is a full duplex stream-oriented protocol. Data is passed to TCP from an upper layer protocol in a continuous fashion. It is then blocked arbitrarily into segments. Being full duplex, the protocol allows data to be sent in either direction between processes.

---

[1] The ones complement of a number is the binary representation with the bits inverted, i.e., 0s replaced by 1s and 1s by 0s.

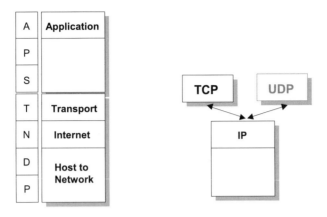

Figure 4.40   TCP

4.12.3.3   TCP Segment

The structure of a TCP segment is shown in Figure 4.41. Each segment serves a dual purpose. It sends data to its peer process at the other end of the transmitting link, and because the protocol is full duplex, acknowledges the receipt of data sent by its peer.

The source and destination port fields are 16 bit numbers local to that host. These fields are combined with the IP address of the host to give the 'socket' number, a unique address called the Transport Service Access Point (TSAP).

In order to know what data has been received correctly, TCP provides sequence numbering. Since the segments can vary in size, this numbering is done on the basis of individual bytes in the transmitted data stream (modulo 2^{32}) rather than by segment. Each segment carries the sequence number in the second field and the acknowledgement number in the third field. The sequence number is the number of the first byte in the segment. The acknowledgement number is the number of the next byte that that host expects to receive from its peer, and therefore acknowledges to the peer that all data up to that point has been received correctly.

Each segment need not be acknowledged individually. The recommendation is to wait for half a second after receiving a packet before sending an acknowledgement in case further data is received, reducing the load of acknowledgement only segments. If a second segment is received within this time, both are immediately acknowledged. If the half second passes without a further segment, an acknowledgement is sent for the first segment on its own.

Lost segments could result either in data being lost or an acknowledgement being lost. The first case can be identified by the destination acknowledging a lower sequence number than expected. The data then has to be resent. A lost acknowledgement will cause the transmitter to resend the data after a time-out. If the receiver receives data it has already received correctly, it sends an additional acknowledgement and discards the duplicate data. In fact, an acknowledgement for a subsequent segment may be received within the time-out. Since acknowledgements specify that all data up to the given byte has been received, this may also acknowledge the data for which the original acknowledgement was lost, avoiding the need to resend the data.

TCP allows various optional elements in the header to allow for negotiating session parameters. A 4-bit header length field indicates the length of the header in 32 bit words. Usually the options field is not used, resulting in a header length of 20 bytes.

The flag field contains six flags as follows:

- **URG**: Indicates that the Urgent Pointer field is to be used. The latter contains the number of the byte in the data field where the urgent data ends. Although this is defined, many implementations do not make use of this feature.
- **ACK**: indicates that an acknowledgement is being carried in this PDU and thus the Acknowledgement Number field is carrying a valid number.
- **PSH**: Indicates that data the peer holds awaiting transmission should be sent on directly and not stored in a buffer (i.e. pushed).
- **RST**: Used to indicate a connection should be reset.
- **SYN**: Used to establish a connection.
- **FIN**: Used to terminate a connection.

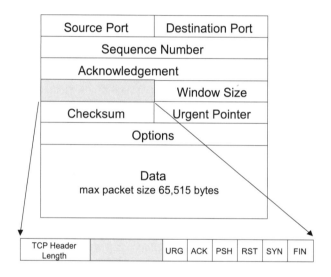

Figure 4.41   TCP segment

The window size field indicates the number of bytes that the receiver would be able to accept.

TCP has the same issue of a lack of receiver address as UDP, which is avoided it in the same way by constructing a pseudo-header with the source address, destination address, protocol and length of the IP packet header which is used with the other 16 bit words in the segment to construct the frame check sequence. However, the frame check sequence for TCP is not optional as it was in UDP. Like UDP, this restricts TCP to working over IP rather than any network protocol, but does reduce the overall overhead of TCP and IP to 40 header bytes.

To see TCP in action, let us return to the example of web page retrieval from Section 1.4, with the web server as Bob and the client as Alice. In order to set up a connection, Alice sends Bob a TCP segment, with the SYN flag set, and a sequence number of Alice's choice (in this case 5637114). The TCP segment does not contain any data. If the connection can be set up, Bob then replies with (a) his own sequence number (in this case 2454387047), and an acknowledgement of Alice's sequence number, which is $5637114 + 1 =$

5637115, because that is the next byte he expects to receive. Since this segment contains an acknowledgement, the ACK flag is set, as well as the SYN flag. Finally, Alice acknowledges Bob's acknowledgement, sending no data, but an acknowledgement number of 2454387048 (2454387047 + 1).

Two other things happen. All these segments contain window sizes specifying the amount of data that they are willing to accept back. Secondly, the maximum segment size (MSS) is negotiated through optional fields in the TCP header. Alice asks for a MSS of 536, as does Bob.

The sequence is as follows:

1. Alice sends Bob a segment with sequence number = 5637114, no data, no acknowledgment, a window size of 8192, SYN set and a MSS of 536
2. Bob sends Alice a segment with sequence number = 2454387047, no data, an acknowledgement of 5637115, a window size of 9112, SYN set and a MSS of 536
3. Alice sends Bob a segment with sequence number = 5637114, no data, an acknowledgement of 2454387048, and a window size of 8576

The connection is now established.

Alice now sends Bob a segment. This segment contains the web page request. Normalising the sequence numbers to 0, Alice sends 285 bytes of data, with a sequence number of 1 (the number of the first byte of data in the segment) and an acknowledgement of 1 (because Alice expects the next byte from Bob to be byte 1). In this instance, the PSH flag has also been sent to request that the data be delivered to the application immediately. When PSH is used is not defined in the standard but is up to the specific implementation on a given computer. Bob replies with no data, but an acknowledgement of 286 (285+1). Bob then sends his own 536 bytes of data, the first part of the web page. Since the web page is more than the MSS in length, Bob splits it into blocks each one MSS in length, until the last one, which is only 450 bytes. Alice acknowledges these transmissions, sometimes sending one acknowledgement for two packets if these packets are received within 500ms. The complete sequence is as follows:

```
 1   Alice → Bob   PSH Seq = 1,     285 data bytes, Ack = 1 win 8576
 2   Bob → Alice   ... Seq = 1,      no data bytes, Ack = 286 win 9112
 3   Bob → Alice   PSH Seq = 1,     536 data bytes, Ack = 286 win 9112
 4   Alice → Bob   ... Seq = 286,    no data bytes, Ack = 537 win 8576
 5   Bob → Alice   ... Seq = 537,   536 data bytes, Ack = 286 win 9112
 6   Bob → Alice   PSH Seq = 1073,  536 data bytes, Ack = 286 win 9112
 7   Alice → Bob   ... Seq = 286,    no data bytes, Ack = 1609 win 8576
 8   Bob → Alice   ... Seq = 1609,  536 data bytes, Ack = 286 win 9112
 9   Bob → Alice   ... Seq = 2145,  536 data bytes, Ack = 286 win 9112
10   Alice → Bob   ... Seq = 286,    no data bytes, Ack = 2681 win 8576
11   Bob → Alice   PSH Seq = 2681,  536 data bytes, Ack = 286 win 9112
12   Alice → Bob   ... Seq = 286,    no data bytes, Ack = 3217 win 8576
13   Bob → Alice   PSH Seq = 3217,  450 data bytes, Ack = 286 win 9112
14   Alice → Bob   ... Seq = 286,    no data bytes, Ack = 3667 win 8126
```

4.13   Internet Protocol

IP is the central pillar of the Internet. The Internet Protocol was designed primarily for internetworking as being a simple protocol almost any network could carry. IP provides a 'best effort' service over the network layer in the form of a datagram service. Data from the transport layer (TCP or UDP) is converted into IP datagrams and carried over the network. An IP network is a network of nodes connected using IP, but since those connections may themselves be formed over other networks, the IP network may be considered to be a network overlaid on networks of other protocols used for the actual data transport.

So as to allow wide application to different network types, IP places very few requirements on the underlying network. It considers the network to be both 'dumb' and 'unreliable', so it is left to the end points or applications to confirm that application data has been delivered correctly, i.e., uncorrupted and meeting any timing requirements. If it is not, it is up to them to take any necessary action.

This simplicity extends to the intermediate nodes, or routers, within the IP network. These have two main functions to perform: to route and forward IP packets towards their destination, and to fragment larger blocks into smaller IP blocks should this be required for the underlying transport network technology.

4.13.1   IP Packet Format

The current version of IP is version 4. An IPv4 packet consists of a 13-field variable length header plus the data field itself. The maximum length of an IP datagram, header and data, is 64 kbytes. The header fields subdivide roughly into blocks of 4 bytes, and convention has it that the IP packet is represented as rows of 32 bits, as shown in Figure 4.42.

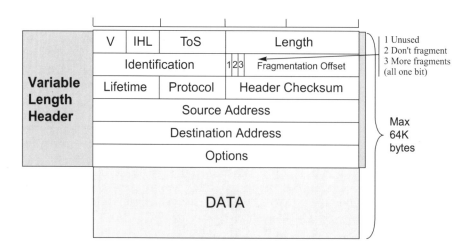

Figure 4.42   IP packet

The first row corresponds to four fields associated with general formatting of the IP packet. The Version (V) field (4 bits) identifies the version of the protocol being used. The Internet Header Length (IHL) field, also 4 bits, indicates the length of the IP header in 32 bit blocks and thus whether any optional header features are invoked after the address fields.

The third field is the 8 bit long Type of Service (ToS) field. In theory, this allows the host to specify priorities (0-7), to indicate whether delay, throughput or reliability are of prime importance, but in practice this tends to be ignored by most routers.

The total length field is fairly straightforward. A16 bit field gives a maximum datagram length of $2^{16} = 65,536$ bytes, but typically datagrams are less than 1500 bytes, and are often limited to 576 bytes, the minimum size all IP transport mechanisms are guaranteed to carry.

The second row within the header is used to manage the process of fragmentation, where larger packets are broken down into smaller packets for transmission. Fragmentation can take place at intermediate routers in order to match the capabilities of the underlying network, but once fragmented, a packet is not reassembled until it reaches the receiver.

The 16 bit identification field allows the host to determine which datagram a fragment belongs to. All parts of a datagram will have the same identifier field.

The 'Don't Fragment (DF)', 'More Fragments (MF)' and 'Fragment Offset' relate to the management of datagram fragmentation. When set, DF indicates that fragmentation of the packet is not allowed. MF indicates that more fragments are following, and all fragments bar the last one have this field set. The fragment offset indicates where in the current datagram the current fragment fits. The field is 13 bits long, 3 bits shorter than the 16 bit IP packet length for the DF and MF bits as well as one which is not currently used, so fragmentation takes place into IP datagrams of integer number of 8 byte blocks.

The process of fragmentation is illustrated in Figure 4.43.

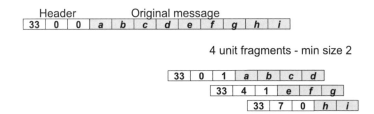

Figure 4.43   Fragmentation of an IP packet

The third row deals with aspects of monitoring a packet's progress through a network. The Lifetime or Time To Live (TTL) field is used to count hops and prevent packets from overstaying in the network, through, for example, circular routing. Originally intended to be a time-based count, it has become hop-based count by default as most routers were unable to give an accurate and co-ordinated version of time. The router simply decrements the field by one each time a packet passes through. When the field reaches zero, the packet is discarded.

The protocol field indicates which transport protocol the datagram is associated with, for example, TCP or UDP. It is fully defined in RFC 1700.

The checksum is used to protect the header. Since the lifetime field is changed by the router, the checksum needs to be recalculated for each hop. It is based on the simple binary addition of successive two byte words and a ones complement of the final result. The checksum is very basic and provides a rudimentary level of protection, although depending on their position, as little as two single bit errors can go unnoticed. Packets with corrupted headers are discarded.

4.13.2   Addressing

A key aspect of IPv4 operation is routing. In order to route packets, it is necessary to know the destination of that packet, and it is also useful to know its source. In IPv4 this is done explicitly via two 32-bit address fields giving over 4 billion addresses. At first glance, this would seem an appropriately large number, but closer inspection as to how these addresses are organised shows a different picture. The structure of IPv4 addresses is shown in Figure 4.44.

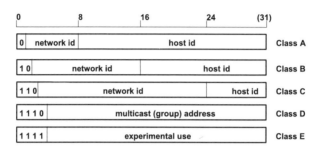

Figure 4.44   Different classes of address

In IPv4, the addresses are defined in terms of 5 possible classes (A, B, C, D and E), although most people think of the addressing in terms of 3 classes as the last two are used for multi-cast operation and for future networks. The class (or group) that an address belongs to is determined by the state of the first, second, third and fourth bits. By convention, IP addresses are hierarchical whereby an address is expressed in terms of the network identifier and a host identifier where the class of an address indicates the type of network and the number of individual devices supported by that network.

IP addresses are 4 octets wide and are normally written in a dotted decimal notation as follows: xxx.xxx.xxx.xxx. The value in the first sub-field defines the format of the inter address.

Class	Range	Format
A	0 - 127	nnn.hhh.hhh.hhh \8
B	128 - 191	nnn.nnn.hhh.hhh \16
C	192 - 223	nnn.nnn.nnn.hhh \24

In all cases, a network ID of all zeros is used for default routing and is not assigned to a particular network. Likewise, a network ID of all ones is used for loopback purposes and is not assigned. Similar limitations apply to the host IDs; all zeros are used to indicate the current network while all ones indicate broadcast packets.

4.13.3   Subnets

Consider the case when an organisation requires to address a number of separate networks. One approach is to assign each segment its own Class B or C address depending upon the number of hosts. A class C address could be used if the number of hosts was certain to be less than 254, otherwise Class B addressing would need to be used. However, this would mean that external routers would have to know which network within the organisation a host is on since they appear as separate networks. The organisation does not have the freedom to switch hosts between networks, restricting any changes that could be made by the organisation's network managers. Also, assigning several networks to an individual organisation is inefficient since a Class B address consists of a block of $2^{16} - 2$ addresses. If the number of terminals on one network was, for example, 300, clearly Class B addressing is required but such an allocation leaves 65234 addresses unused.

One solution to the above problem is the use of subnets. In the IP domain, this refers to the case when a network is split up internally (in addressing terms) into a number of subnetworks, but remains as a single entity when viewed by the outside world, i.e. nodes not within that network. This is done by introducing a third hierarchy into the address structure between the network and the host part of the address, so that effectively part of the host address space over which the network manager has control becomes reserved to describe not a host machine but a subnet within the wider organisation.

External routers direct packets to the network in the normal way based on the network address. Local routers then use a subnet mask to determine which other local router a particular packet is destined for – the local router does not need to know exactly which host unless the host is associated with its subnet. This means that only routing tables of nodes within the network need be changed to reflect the new three (rather than two) level hierarchy that exists. External routers are unaffected by changes within the network. Subneting is defined in RFC 950. A simple example of subneting is shown below where by an organisation's three separate networks are presented to the rest of the network as a single network.

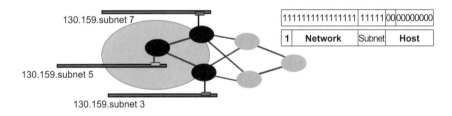

Figure 4.45   Subnet addressing

In the example shwn in Figure 4.45, any packets destined for any of the 130.159/16 LANs arrives at the router associated with 130.159.28/22 segment. The router then simply uses the subnet mask to obtain the destination address subnet via a simple 'and' operation. The packet is then forwarded by the router to the appropriate subnet router. The 130.159/16 Class B network has been subnetted into 64 possible subnetworks each with 1022 hosts.

As subnetting is an internal operation and hidden from the 'rest of the world', the internal configuration of the subnets is left up to the network managers - only the internal routers need to be programmed with the appropriate masks.

For example:

- A mask of 255.255.252/22 gives 64 subnets of 1022 hosts.
- A mask of 255.255.224/19 gives 8 subnets of 8190 hosts.
- A mask of 255.255.255/24 is the subnet equivalent of Class C addressing.
- A mask of 255.255/16 is the subnet equivalent of Class B addressing.

4.13.4   Classless Addressing

The addressing scheme suggested by IP can be rather restrictive and has resulted in rather inefficient use of the address space (which is not large enough now anyway) and a lack of Class B addresses. Classless InterDomain Routing (CIDR), defined in RFC 1519, operates by allocating addresses to networks that originally wanted Class B address (but would not have made use of them all) in blocks of the Class C addresses. Effectively, CIDR no longer recognises the distinction between Class A, B and C addresses – hence the name classless. CIDR allows address assignment no longer to be limited to /8, /16 or /24.

Routers treat these addresses in much the same way subnet routing was applied – associated with each set of classless addresses is an address mask whereby each incoming packet address is masked to get the base address and thus identify the router associated with that network. The size of the address mask can now be variable.

Tied in with CIDR is the concept that particular address ranges are associated with geographical areas; e.g. 194.0.0.0 to 195.255.255.255 are for Europe, 198.0.0.0 to 199.255.255.255 are for North America. This makes routing between networks easier. Route Aggregation is used whereby a single network prefix can be used to describe or represent multiple networks.

Consider the case when an organisation has been assigned a range of IP addresses 130.159.80.00 up to 130.159.95.255, which it will allocate to a number of departments as appropriate. This is illustrated in Figure 4.46.

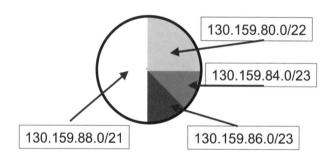

Figure 4.46   Example of CIDR addresses

In the wider context, the router will advertise itself to the network that it should receive any packet whose first 20 bits of address are equivalent to 10000010100111110101, i.e., 130.159.80.0/20. In the above example, four departments require 510, 510, 1022 and 254 hosts and thus the 130.159.80.0/20 space can be assigned as above.

Should one of the organisations change ISP, then CIDR allows them to keep the assigned range. The new ISP would advertise both its own range of IP addresses and the old range of its new customer. For example, consider the user with range 130.84.0/23. The original provider would advertise its range of 130.159.80.0/20, while the new ISP would advertise 130.159.84.0/23. Intermediate routers would use longest prefix matching to recognise that 130.159.84.0/23 packets should be forwarded to the new ISP and not the 130.159.80.0/20; the /23 has precedence over the /20.

4.13.5   Domain Name Service

While IP addresses are efficient from the point of view of computers and routers, they are not particularly user friendly from the point of view of users. For this reason, the numbers are replaced with text to form *domain names*. Generally speaking, domain names are hierarchical, consisting of a sequence of elements separated by periods which, from right to left, identify first the network are then the individual host, and sometimes the service (www, ftp, etc.) on a host. The right-hand element is the *top level domain,* which is either based on geographical location (uk, us, au (Australia, for example) or type of organisation. The latter are mainly American, although to get the '.com' registration, many businesses register their domain names in America.

Mapping from domain names to IP address is done using data bases holding the IP address for each domain name. The databases are held by *name servers*, access through the Domain Name Service (DNS) protocol, and the system is hierarchical. Top level domains are registered with one of a number of registries, the main one being InterNIC in the United States, although the system was changed recently to allow some competition from other name server providers. Top level domains are split into subdomains, and the registrant of the top level domain registers the location of name servers for each of these domains. The registrant then has the ability to add new names within these subdomains without contacting the top level nameserver. For example, the 'uk' top level domain is subdivided into ac.uk, co.uk, gov.uk and org.uk, and the ac.uk subdomain, for academic institutions, is run by JANET. The subdomain strath.ac.uk is assigned to Strathclyde University, who register their name servers with Janet. This means that they can assign the subdomain eee to the Electronic Engineering department, and garland to an individual host within that network, while only updating their local name servers (see Figure 4.47).

In order to have short, memorable, addresses, many companies register directly with a top level domain, and then arrange for requests www.organisation.com to go to a web server for their company. The actual domain registered is organisation.com, and since the company can arrange for any IP address to be matched to that domain name, the web server can be located on any host and network, perhaps even one run to another company which specialises in hosting web servers.

DNS also provides additional services such a host aliasing, mail server aliasing and load distribution, by varying the IP addresses returned. It is formally specified in RFC 1034/1035.

IP address and host name mappings are stored in the form of resource records (RR). These consist of 4 elements: name, value, type and time to live. There are four main types of record. If the name server knows the IP address of the requested domain name, this is stored in a type A (authoritative) record, which matches IP addresses (value) to hostname (name). If the name server does not know the IP address, it will know the address of another nameserver which should know (or itself know one that will), and type NS records maps a domain name (name) to a Name Server (value). Type MX records maps the host name of a mail server (name) to its alias (value).

The final record is a type CN record which maps a canonical name (Value) to a hostname (Name). Some computers are given additional, usually shorter, names for convenience, and the CN record maps these names on to the full hostname. The ITR computer in the example in Section 1.4 answers to the name of www.itr.unisa.edu.au, although its full (i.e. canonical) name is charli.levels.unisa.edu.au. This system makes web hosting, for example, easy. Anyone can purchase a domain name and register it, and make an arrangement with a company to host

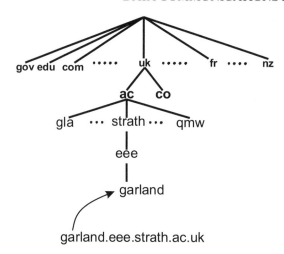

Figure 4.47 Naming hierarchy

their pages. CN records then point their domain name at the hosting company's computers, so the domain name owner does not need to have a computer.

Local servers store previous results of DNS requests in their cache. This enables the DNS process to be speeded up as there is then no need to contact other servers. Clearly, there is a balance to be struck between retaining and discarding cache entries. The final field of the record, time to live, indicates the time the record should remain in the cache before expiring.

DNS messages are exchanged between hosts and name servers. It works by using UDP to call appropriate name servers to map names to IP addresses. DNS returns the resource record held by the server which contains the IP address so that application, e.g. telnet, ftp, http, can then use it.

Name servers can be Local, Root and Authoritative. Intermediate name servers may also be used in the communication chain.

DNS queries may be either recursive or iterative. In the former (the most common), each domain server obtains the mapping on behalf of the server or host who made the request. Iterative queries (typically between root and local to minimise processing load at root) result in either the required mapping or the address of the next server in the chain which then allows the next DNS query to be made directly. In the example in Figure 4.48, (2) is an iterative query which yields the response (3) which tells the local server to contact the intermediate server directly. If (2) was recursive, then (3) would be directly to the intermediate server and there would be NO direct dialogue in any direction between intermediate and local servers.

Figure 4.49 illustrates the sequence of events for setting up a TCP connection to host www.dufc.co.uk

1. DNS query for host www.dufc.co.uk
2. ARP request for 130.159.86.4, ARP reply from 130.159.86.4
3. DNS query sent to 130.159.86.4 (resolved) If the request is not resolved locally, further requests are made to additional non-local DNS servers
4. DNS reply from 130.159.86.4
5. Forwarding decision based on default entry in forwarding table
6. ARP request for 130.159.86.1, ARP reply from 130.159.86.1

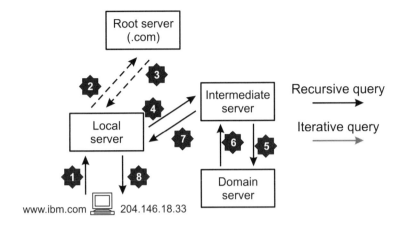

Figure 4.48   Recursive and iterative DNS queries

7. TCP connection request sent to www.dufc.co.uk via 130.159.86.1
8. TCP connection request arrives at www.dufc.co.uk,
9. TCP connection reply returned
10. TCP connection reply arrives at garland,
11. HTTP request sent to www.dufc.co.uk

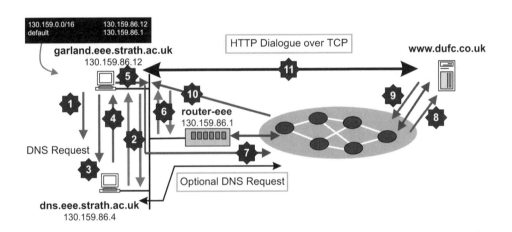

Figure 4.49   Chain of events for a DNS query to www.dufc.co.uk

4.13.6   IP Routing

Routing in IP networks is really just like routing any data network as the same basic principles apply. The network is considered as an interconnection of so-called 'autonomous networks'. Each of these autonomous networks is self-contained and can run its own routing algorithm. The issue is then the interconnection of these different networks, which requires an agreed Internet routing protocol between neighbouring networks. There are now two common protocols, OSPF and BGP, replacing the original Routing Information Protocol (RIP) and Exterior Gateway Protocol (EGP) used on the Internet. RIP and OSPF are designed for routing within the individual autonomous networks, while EGP and BGP refer to routing between autonomous sub-networks.

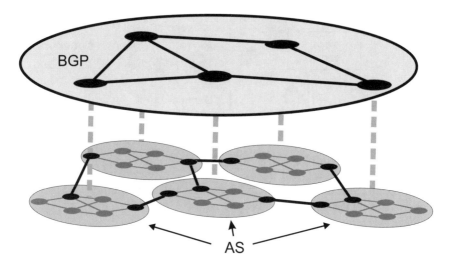

Figure 4.50   Routing between autonomous networks

- **RIP** – Routing Information Protocol. Intra-domain protocol based upon distance vector routing. Based upon a hop-count (limited to 15) and updates every 30 seconds.
- **OSPF** – Open Shortest Path First routing applies to the problems of routing within each sub-network. Essentially, it is a version of link state routing discussed earlier. Allows authentication of routing messages and multiple routes. Also enables a hierarchy to set up whereby domains are divided into areas so that packets do not need always to be forwarded to exact domain – just in the right direction.
- **BGP** – Border Gateway Protocol. This is essentially a protocol that defines how 'reachability' information is exchanged between routers. Traffic can terminate at an autonomous network or pass through (transit) to reach its destination network. BGP is not concerned with finding the 'best' path – any path will do, as long as it does not loop back on itself. Packets will pass through different domains using different routing regimes so optimal routes are not really possible. There is some element of choice so 'good' paths can be found. The selection policies can be based upon technical, business, political, cost issues. This is left up to individual BGP routers. The complexity of BGP grows as the number of autonomous networks grows, rather than as the number of hosts within these networks grows. The complexity of intra-domain routing protocols like RIP and OSPF will depend on the number of hosts or sub-networks within each autonomous network.

4.13.7 IPv6

IPv6 is the next generation of IP. An initial spur to its development was a perceived lack of address space for Ipv4, as the number of hosts and networks grew. In fact, dynamic allocation of IP addresses by Internet Service Providers and classless addressing have extended the life of the address space of Ipv4. New developments in mobile phone technology and ASDL, both of which will have 'always on' connections, will require additional IP addresses, and although there are work-arounds to allow IP addresses to be shared within a network and translated by the gateway into a smaller dynamic range of IP addresses and ports visible to the external network, the IPv4 address space will run out sooner or later.

IPv6 takes no chances with addresses running out. The new addresses are 128 bits long. With this size, as well as being able to given every person, terminal, and even light switch an IP address, electronic devices will be able to be assigned addresses without any concern about these addresses being relinquished at the end of their life, in a similar way to Ethernet card addresses. The additional addresses also allow multiple addresses per device, perhaps for different services.

An additional feature is the introduction of an 'anycast' address, so that a packet is sent to any one of a number of terminals. This would allow the easy addressing of requests for services which are provided in a number of places so that the local provider is used.

An additional address space is not the only advantage of IPv6. Many services which were never envisaged to be transmitted over a connectionless IP protocol are now being used, and the lack of Quality of Service support in IPv4 has been a significant problem. Ipv6 includes a traffic class and a flow label which would allow packets from a particular flow to be identified for special handling, should additional resources, for example, be negotiated in advance. IPv6 also has improved security support, and a mechanism for adding optional fields to the header for expansion.

Fragmentation is handled slightly differently in IPv6 compared to IPv4. The minimum packet size which must be handled is increased from 576 bytes for IPv4 to 1280 bytes for IPv6. Also, IPv6 routers do not fragment packets if required, but rather send a message back to the transmitter to ask it to do it. This means that routers can be simpler. IPv6 also has the capability of using larger packets called jumbograms.

The simplification of routers is assisted by that fact that IPv6 headers have fewer fields than IPv4 headers. However, even without any optional fields they are twice the size of an IPv4 header. For protocols that generate large numbers of small packets, like Voice over IP, this is unfortunate, but the additional quality of service support mitigates this. However, IPv6 has not so far received wholehearted support amongst manufacturers and it may be some time before it gains widespread acceptance. Part of the problem is that while it is relatively easy to provide user terminals with dual stacks capable of using IPv4 or IPv6 depending on what they are communicating with, if IPv6 is not supported on the core network, or indeed on any link between the parties, the IPv6 packets have to be carried over IPv4 encapsulated in IPv4 datagrams. This process is known as tunnelling. This adds additional overheads, and because IPv4 does not support the quality of service provisions of IPv6, the quality benefits will not be obtained. This means users will get little for their investment until a large number of others also upgrade their equipment – a chicken and egg situation.

The IPv6 packet is shown in Figure 4.51. The V field gives the IP version. The P (Priority) allows different service types with different QoS requirements. P = 0 – 7 flow controllable data, i.e., data that is not real-time, while P = 8 – 15 indicates real-time data which is not flow controllable data. The Flow Label is intended to allow nodes to differentiate between traffic flows and their demands. Along with Source and Destination addresses, the flow label is used to uniquely address a 'flow'. Length is a 16 bit field which defines the length of the payload in bytes. The Next Header field indicates that optional extension headers are used, i.e. additional functions are associated with a packet. Six types of extension defined: Hop-by-hop options, Routing, Fragmentation, Authentication, Encryption, Destination. These optional extension headers are used only when required and in a particular order. Optional headers have fixed format or are specified in terms of Type, Length and Value fields. If Next Header in packet or extension matches the payload field of IPv4, then no additional functionality is demanded. Finally, the Hop Limit field decrements as a packet travels across a hop; at 0 the packet is discarded. It is similar to the TTL field in IPv4.

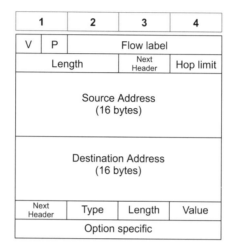

Figure 4.51   IPv6 frame structure

4.14   QoS over IP

The TCP/IP stack offers, essentially, a 'best effort' service. Such a service does not really match the diverse needs of many applications that require much tighter constraints on the delivery of information. A number of mechanisms have been proposed as means to provide a QoS aspect to IP transmission.

4.14.1   Integrated Services

Integrated Services (Int Serv) attempts to allocate resources to uni- and multi-cast traffic flows on a flow-by-flow basis. Int Serv requires the network to have the ability (via its routers) to allocate resources to particular flows (source/destination pair) and the ability at various points to monitor or police flows or aggregated flows. This is a fundamental change to the standard Internet model which dealt with flows purely on an end-to-end basis only. In the Int Serv model routers are required to identify streams and keep a note of their current state. It requires an explicit mapping or binding between flows and services. Key elements within Int Serv include admission control, reservation protocol and policing along with routing and packet forwarding mechanisms.

4.14.2   Resource Reservation Protocol

A key part of Int Serv, and indeed other QoS related Internet schemes is a Resource Reservation Protocol – RSVP.

Essentially, this protocol is a mechanism by which resources are reserved along a previously obtained path for either uni-cast or multi-cast communications. RSVP is receiver-oriented whereby requests for resources come from the destination rather than the source.

The first stage in the protocol is the setting up of a path between source and destination. RSVP achieves this by sending special IP packets known as PATH messages which allow routers between source and destination to note the 'reverse path'. Once the receiver receives the PATH message (which also contains the traffic spec), it returns a RESV message back along the router path (see Figure 4.52). The RESV message contains the QoS specifications for both the receiver and transmitter. As these messages propagate back along the path, each router works out to see if a reservation can be made that meets the traffic requirements of each party.

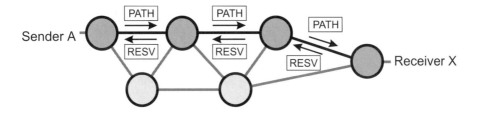

Figure 4.52   RSVP operation

If such a reservation can be made, i.e. the router feels it has sufficient resources to meet the QoS requirements, it passes the RESV message back to the next router in the tree. This process is continued until the RESV message gets to the source or, in the case of multicast, it gets to the first router that supports that particular stream. If at any point a router decides that it cannot support such a flow and its QoS constraints, then an error message is sent back to the receiver.

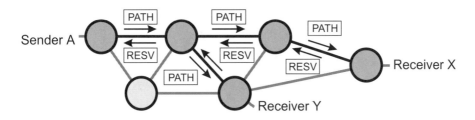

Figure 4.53   RSVP multi-flow operation

Each router maintains a record of the state of the flows that it supports. These states need to be updated regularly. In RSVP, RESV messages are sent by the receivers every 30 seconds or so to maintain that 'connection'. If a router does not receive an update it deletes that record. This helps maintain robustness and also enhance flexibility. Receivers can attempt to change their demands at any point and change their routes without explicitly telling all routers within the system. Should any part of the path fail, IP routing is robust enough to find an alternative route and RSVP connections can react automatically with minimal effort. Note that the receiver has no knowledge of the actual path the flow actually takes.

RSVP supports multicast connections as flows can be merged and different QoS demands supported (see Figure 4.53). Even multiple senders can be supported.

4.14.3   Differentiated Services

Although Int Serv and RSVP go some way to meeting QoS demands in IP networks, there are a number of major disadvantages. The mapping of QoS demands, i.e. reservations, to specific flows and the associated per-flow processing at each router, is a significant complication, and therefore such an approach is inherently unscalable. RSVP is not always available at hosts and this approach requires quite a complex mapping of QoS demands and network resources.

Differentiated Services (Diff Serv) takes an alternative approach to QoS provision. Diff Serv classifies packets into a small number of clearly identifiable streams which receive a particular service (and allocated resources). These streams, which are aggregated flows, are then handled by Diff Serv routers in different ways as befitting their specification. This approach requires much less effort on the part of the router than the management of individual flows, and therefore is scalable. However, it is not quite as flexible as Int Serv, as flows have to be assigned to a limited number of flow types, but service definitions are not fixed to specific flow types and can be updated and changed.

A service level agreement (SLA) is agreed between source and service provider, and then the service provider maps source stream to a particular stream type that meets QoS requirements. Flows from many different sources are aggregated to a particular stream type according to resources available within network.

Diff Serv has three key elements: packet classification, traffic conditioning and packet scheduling. The first two are performed at the edge of the network, i.e. at the first Diff Serv router the IP packet meets. The simpler packet scheduling function (i.e. forwarding) is done by intermediate routers within the core network.

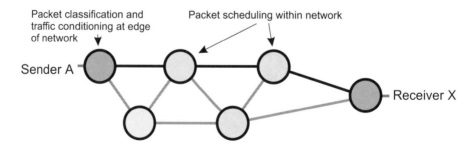

Figure 4.54   Diff Serv functionality

Packet classification is, essentially, the mapping of a user's flow on to a particular stream type. The edge router determines which is the appropriate flow by considering the SLA (i.e. QoS requirements) and the streams available. Packets from this user's flow will then be marked to indicate they are mapped to a particular stream.

Traffic conditioning is also applied at the edge of the network. This consists of metering the incoming flow to ensure that it meets the agreed traffic profile. Packets falling outside the profile are either re-marked to a lower service class stream or traffic shaping is applied to conform with the profile.

Diff Serv routing is still IP routing, the packet scheduling function merely determines the mechanisms by which packets are served and not the path they take over the network. In Diff Serv, an 8 bit field in the IP packet is used to identify exactly what type of service an IP packet should receive from a DS router, i.e. the Per Hop Behaviour (PHB). In effect PHB represents a class of service. It does not define how such behaviour is achieved – just what it is.

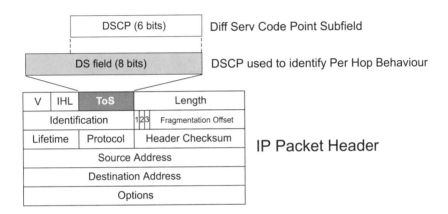

Figure 4.55   Diff Serv identification

Six bits of the Diff Serv field are used – the Diff Serv CodePoint subfield – giving 64 possible behaviour types. The remaining two bits do not specify any PHB but do ensure a degree of backward compatibility with non-Diff Serv routers. The proposed means to carry the DS field are the ToS field in IPv4 (see Figure 4.55) and the Traffic Class field in IPv6. The latter is a union of part of the priority and flow label fields.

In standard Diff Serv operation, the Diff Serv field is set at the edge of the network and ideally remains constant as the IP packet travels to its destination. However, it is possible to change or re-mark this field should the packet be 'out of profile' or the PHB and Diff Serv fields need to be re-mapped (when travelling through new Diff Serv domains).

4.14.4   Multi-Protocol Label Switching

IP routing is based upon autonomous shortest path routing which can conserve network resources but can lead to congestion on some links and under-utilisation on others.

MPLS allows packets to be routed based upon a specific label rather than the IP addresses within the IP packet header. Specific devices known as Label Switched Routers (LSR) are used to do this. LSR matches the label with an entry in its forwarding table and passes the packet to the appropriate output link, having replaced or updated the packet's label as appropriate. IP packets are assigned labels as they enter the MPLS domain (by the ingress router) and have it removed as they leave the MPLS domain (by the egress router) (see Figure 4.56). Packets follow a labelled switched path (LSP) which is determined prior to transmission and chosen to

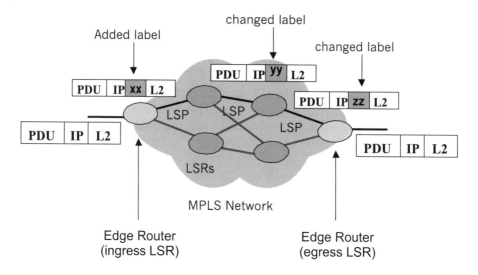

Figure 4.56   MPLS operation

match the QoS of the source or receiver. LSPs can be determined using a variety of schemes including RSVP, and constraint-based routing.

4.15   Questions on the Network Perspective

4.15.1   Questions on Network Configuration

1. A communications network consists of N nodes. For the following cases, determine the number of links required to connect these nodes assuming that each link is bi-directional:

   (a) A 5 node ring
   (b) An 8 node fully interconnected mesh
   (c) A 5-connected, 10 node partially interconnected mesh
   (d) Repeat the above three cases when $N = 96$.

2. Determine the connectivity of the two networks given in the following two diagrams. For both networks, how many iterations are required to determine the connectivity?

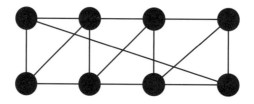

	A	B	C	D	E	F	G	H	I	J
A		1		1	1					
B	1		1		1	1				
C		1		1		1	1			
D	1		1					1		
E	1	1							1	
F		1	1				1		1	
G			1			1		1		1
H				1			1			1
I					1	1				1
J							1	1	1	

4.15.2   Questions on Switching Techniques

1. A 64 kbit/s circuit-switched path between two users utilising 9 switching stages takes 5 seconds to be set up. How long will it take, in total, to directly transfer the entire contents of a 1.4 Mbyte floppy disk between the two users using this channel?
2. If the above circuit switched path is replaced by a 9 node-packet switched system utilising 1 Mbits/sec links, how long does it take to transfer the 1.4 Mbyte file, assuming that there is no delay at any of the switching nodes and that packet payload is 256 octets and packet overheads are equivalent to 18 octets?
3. Contrast the delays when packet payload is changed to 128, 512, 1024 and 4096 octets respectively.

4.15.3   Questions on Network Dimensioning

1. A simple telephone exchange consists of 4 circuits. If, at the busiest time of day, calls arrive at the exchange every 240 seconds, calculate the occupancy of the exchange and determine the probability that a call will be rejected.
2. If, in the above system, the users are happy with a loss of 2%, determine the number of circuits required in the multiplexor.
3. If the call rate in the above system increases fivefold, determine the call loss probability and utilisation of each of the 4 exchange circuits.
4. A terminal concentrator consists of six 56 kbit/s input lines and a single 128 kbit/s output line. The mean packet size is 450 bytes and the arrival rate associated with each input line is 5 packets/sec. What is the mean delay experienced by a packet and what is the mean number of packets stored in the concentrator?
5. Two computers are inter-connected via a 64-kbit/sec line and currently support 8 interactive sessions (connections). If the mean packet length is 150 bits and the arrival rate/session is 4 packet/s, should the network provide each session with its own dedicated 8 kbit/s channel

or should all sessions compete for the entire line capacity when packet delay is the most important criterion?

4.15.4   Questions on Routing

1. Routing within a 6 node network is based upon a distance vector approach; router A is adjacent to routers B, D and E. The distance vectors associated with each of these 3 routers are given below. Using these tables, derive a new routing table for node A. Assume that the current distance between node A and its neighbours is accurately reflected in the tables.

NODE B			NODE D			NODE E	
A	10		A	20		A	15
B	-		B	15		B	10
C	20		C	40		C	25
D	30		D	-		D	10
E	10		E	10		E	-
F	10		F	25		F	5

2. The current link state tables for a 6 node network are given below. Draw this network and determine the routing table associated with node A.

Node				
A	B(10)	D(5)	E(20)	
B	A(10)	C(20)	E(10)	F(10)
C	B(20)	F(20)		
D	A(5)	E(10)		
E	A(20)	B(10)	D(10)	F(10)
F	B(10)	C(20)	E(10)	

3. Router A is required to broadcast a packet to all other routers in the network above. If the packet field contains a hop count initialised to the diameter of the network, determine the minimum and maximum number of packets this network may be required to transmit in order to convey this broadcast packet.

4.15.5   Questions on Error Control

1. A data link protocol has the following characteristics:

Data length	100 bytes
Header length	8 bytes
Channel capacity	2 Mbit/s
Acknowledgement frame length	8 bytes
Probability of lost frame	2×10^{-3}
Probability of lost acknowledgement	5×10^{-4}
Service and propagation delay	0.15 ms

(a) Estimate the maximum possible efficiency of this system if the protocol operates in a simple 'stop and wait' manner using positive acknowledgements. What is the impact of setting a timeout interval to be 1 ms?

(b) In order to maximise link efficiency, the 'stop and wait' protocol is to be changed into a sliding window protocol. Estimate an appropriate window size, if in the first instance the link is assumed to be error free.

2. The key parameters of a data link and its ARQ protocol are given in the table below. If a 3-bit SEQ field is employed, which retransmission mechanism gives optimal link performance?

Information field	150 bytes
Frame overhead	16 bytes
Acknowledgement length	16 bytes
Probability of lost frame	10^{-3}
Probability of lost acknowledgement	2×10^{-4}
SEQ field size	3 bits
Acknowledgement time out	150 ms
Link bit rate	64 kbit/s
Link propagation delay	50 ms

3. A dual speed point-to-point communication link utilises a dual mode data link protocol that is capable of being switched between a forward error control mechanism or an ARQ protocol based upon a go back N retransmission scheme. This link operates in one of two speeds – low and high – and has key characteristics as shown in the table below. The link's FEC mechanism is characterised by a 50% redundancy and perfect error correction. Which error control mode should be applied to each link speed if the key aim is to maximise link efficiency?

Information field	200 bytes
Frame overhead	16 bytes
Acknowledgement length	20 bytes
Probability of lost frame	10^{-3}
Probability of lost acknowledgement	2×10^{-4}
Window size	8
Acknowledgement time out	250 ms
Low link bit rate	32 kbit/s
High link bit rate	256 kbit/s
Link propagation delay	100 ms

5

The Link Perspective

5.1 Data Link Layer

The data link layer provides the network layer with a connection between two terminals. It has a number of functions, the main ones being:

- **Framing**, so that the different parts of the information and control signalling can be recognised.
- **Medium access**, so that data can be sent over the link as required. Note that this function only applies to shared media – if there is a dedicated transmission line between the communicating terminals no access control is required.
- **Error control**. The data link layer tries to provide reliable data to the network layer, in that it will either be correct or if erroneous, it will be marked as such. This requires the use error correction and detection, although if the information is corrupt, it will usually be left to the upper layers to deal with by ARQ or concealment.

In this chapter, we will look at these functions, and then at some LAN and MAN systems which provide most of the data link functionality of current networks. One of these, ATM, was designed both for network and data link layer functionality. However, the rise of IP is such that that is becoming the dominant technology in the network layer, and since ATM is often used to carry IP, it is discussed here.

5.2 Frame Delimiting

The start and end of a frame must be marked. This is particularly true when the transmission medium is shared, but even if a dedicated transmission line is available, some form of synchronisation is still required. The first possibility is to use a field in the header to specify the number of characters in each frame. The data link layer at the destination reads this field and thus knows where the end of a frame will occur. Such a technique is not robust, and frame integrity would be destroyed by any transmission errors.

A second technique involves marking the start and end of each frame with an ASCII character sequence. If the frame gets corrupted in any way (including start/stop characters) the data link can re-align itself by seeking out the next frame markers, resulting only in the loss of a finite number of frames. It is possible that these sequences of ASCII characters might occur naturally within the frame. To prevent any erroneous interpretation of frame boundaries, specific characters are inserted before each non-frame-boundary character sequence to indicate that these characters do not represent either the start or end of a frame.

This is known as character-stuffing. This scheme is closely tied to 8-bit characters (particularly ASCII codes) and as networks have evolved, the dependence upon 8-bit characters has decreased, rendering this method increasingly unsuitable.

An alternative is to mark the frame boundaries with a unique bit sequence. The most common sequence is 01111110, which is used, for example, in HDLC 5.9.1. To ensure its uniqueness whenever a transmitter encounters 5 consecutive 1's within a frame, it automatically inserts a 0 into the outgoing stream, so 1010101111011011111100101 would become 101010111101101111110100101, for example. This process is called bit-stuffing. The processes of bit or character stuffing are transparent to the network layer as the receiver will automatically remove the stuffed bit/character.

The final method of marking the start/end of a data link frame involves using the physical medium and violating specific coding rules at frame boundaries. This technique depends upon there being some redundancy in the encoded signal and is widely used in LANs and the ISDN S-bus specification. The latter mechanisms are the most common, but regardless of the method, marking frame boundaries will always introduce an element of overhead.

5.3 Medium Access Control

The advantages of statistical multiplexing mean that sharing a common transmission medium is an efficient strategy. However, it brings the additional problem of giving terminals permission to use that medium when they need it, given the use of other terminals. This is the task of medium access control.

There are a number of requirements on any medium access control scheme. These are:

- That the system should be reliable.
- That the protocol should be stable.
- It should given fair access to the system.
- It should be possible to control access to the system according to the needs of the service or for other reasons.
- It should be efficient.
- It should be easy to manage.

A number of different systems are possible. These can be broadly categorised as centralised schemes, where special nodes allocate resources on the network, and distributed schemes, where all nodes operate cooperatively and in a similar manner. Another distinction is between reservation-based systems, where terminals are allocated a particular resource for their use, and contention-based systems, where terminals request resources when they have need of it. Hybrid schemes, incorporating some reservation and some contention, perhaps based on the service required, are also possible. A centralised scheme may use a system whereby contention is used by a terminal to inform the central controller that resources are required, and then reservation is used by that central control. Many radio systems use this strategy.

5.3.1 Contention-Based Schemes

5.3.1.1 ALOHA

ALOHA is the simplest possible contention-based access system. In this system, when a terminal has something to transmit, it transmits it. If another terminal transmits during this time, it will interfere and both transmissions will be corrupted, as shown in Figure 5.1. This is called a collision. If the terminal receives an acknowledgement to its transmission, the data has got to the receiving node. If not, it must have collided with another transmission and failed to be received. The transmitter then waits for a random period of time and then tries again. The fact that this period of time is random is important. If the terminal waited a fixed time, and a collision occurred with another terminal on the first attempt, it would occur on all subsequent attempts as well. The only other sophistication to the ALOHA scheme is to limit the amount of data which is transmitted at any one time. This is not only to be fair to other users and not 'hog' the channel. Since terminals do not listen to see if the transmission medium is free before transmitting, the longer a packet lasts, the more chance another terminal will transmit and interfere with it.

ALOHA is simple, and since it does not require any centralised control, is very easy to manage. However, it is highly inefficient. At best, an average of only 18% of the channel capacity can be used.

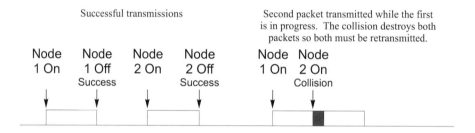

Figure 5.1   Basic ALOHA scheme

5.3.1.2   Slotted ALOHA

While ALOHA is inefficient, the obvious remedy of listening before starting to transmit is not always available. For example, in a mobile radio system, terminals may be communicating with a base station, and due to obstructions or distance may not be aware of other terminals transmitting, even though the signal of these other terminals will reach the base station and interfere with what they are transmitting (see Figure 5.2). This is called the hidden terminal problem.

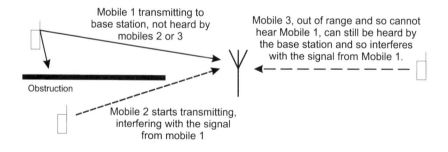

Figure 5.2   Hidden terminal problem

Slotted ALOHA improves on simple ALOHA by constraining the transmissions to occur in frames. This halves the chance of collisions, because while two terminals which decide to transmit in the same frame interval will start to transmit in the next frame (and therefore collide), any terminal deciding to transmit in the second frame interval would not start to transmit until the beginning of the following frame and would not therefore collide. This raises the throughput to a maximum of 36% of the channel capacity. Slotted ALOHA forms part of the medium access control schemes of most mobile radio systems.

5.3.1.3   Carrier Sense Multiple Access with Collision Detection (CSMA/CD)

The CSMA/CD access protocol is used for Ethernet LANs, among others. It is the protocol polite people use for a telephone conference, when the usual visual clues as to when to get a word in edgeways are absent.

- Listen before talking.
- Don't talk while someone else is talking.
- Don't ramble on incessantly.
- Stop and back off if you find yourself starting to speak at the same time as someone else.

The main difference from ALOHA is the fact that nodes listen to see if the link is free before transmitting. This allows much higher throughputs. If the link is busy, the terminal waits until it becomes free. Since a number of terminals may be waiting like this, some implementations of CSMA/CD require that terminals wait for a random interval after the channel becomes free before starting to transmit to reduce the chance of collisions. However, Ethernet uses the common 1-persistent technique, which is to transmit immediately the channel becomes free without waiting.

If two nodes attempt to transmit at the same time; or almost at the same time, then the two frames will collide. Collision detection is achieved by each node monitoring the state of the bus while transmitting a frame. If the data on the link is different from the data being transmitted, then a collision has occurred.

Collisions occur because, although nodes will not attempt to transmit while the link is busy, it takes a finite time for a frame to propagate the length of the bus, so it might start transmitting when a previously transmitted frame is on its way. This is shown in Figure 5.3. The transmission will propagate along the bus at about 2×10^8 ms^{-1}. In the worst case, the colliding terminals will be at opposite ends of the bus (Figure 5.3 (a)), and the second terminal will start to transmit just before the first frame reaches it (Figure 5.3 (b)). Each terminal must continue to transmit to detect that a collision has taken place, otherwise the packet could collide without the transmitter being aware of the fact. The second terminal detects the collision very quickly (Figure 5.3 (c)), but in order for the first terminal to detect it, the signal must propagate back along the bus ((Figure 5.3 (d)). This means there must be a minimum packet length on a CMSA/CD system equal to twice the propagation time over the bus length (plus some additional time for processing). For a 2.5 km coaxial network this time is 51.2 μs, which with a bit-rate of 10 Mbit/s means a minimum frame size of 64 octets.

It is this factor which constrains the capacity of CSMA/CD schemes. If the propagation delay is small compared to the bit rate on the bus, frames can be small, and throughput maximised. However, the propagation time limitation becomes significant as bit rates increase, since either the frame size has to increase, constraining throughput, or the bus length has to be limited. The former solution is often undesirable, not just from the point of view of

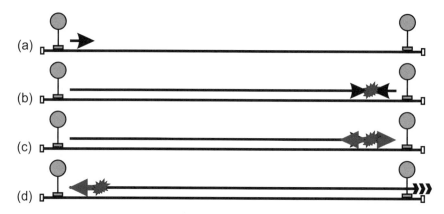

Figure 5.3   Collisions in a bus-based Ethernet system

inefficiency when transmitting small data packets, but because of backward compatibility with previous (slower) versions of the standard. Ethernet gets round the problem in faster versions by using a logical bus over a physical star rather than a physical bus. The hub at the centre of the star can detect a collision by noting if more than one of its connections is active at one time. The hub can then generate a collision signal. This means that the limit is on the segment length to the hub rather than an end-to-end limit on the bus.

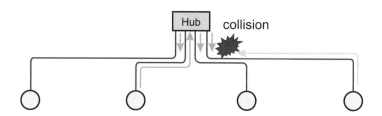

Figure 5.4   Collision detection in a hub-based Ethernet

When a collision occurs, both nodes stop transmitting and will attempt to retransmit the frame after a certain interval known as the 'Back-Off' time. Ideally, the Back-Off time should be different for each node involved in a collision and so involve a random process.

As an example, an algorithm known as the Binary Exponential Back-off (BEB) algorithm used for Ethernet is designed to resolve contention between (up to) 1024 nodes. In the BEB algorithm, after a collision, a node waits X units of time ($51.2\mu s$) where X is a random number between 0 and ($2^N - 1$) for $0 \le N \le 10$, or 0 and 1023 for $11 \le N \le 16$, where N is the number of retransmission attempts. After 16 retransmission attempts, the frame is considered lost and it is left to the higher layers to deal with it.

5.3.2   Reservation-Based Schemes

Reservation-based schemes make more efficient use of the transmission medium because access is controlled and collisions do not occur. These can be categorised as polling systems using a central manager and distributed systems.

5.3.2.1 Polling

In a polling scheme a central controller is responsible for permitting access to the shared medium. It issues messages to the hosts permitting them to use the medium for a certain period of time.

The simplest method is a 'round robin' scheme, where each host is contacted – polled – and invited to transmit its data. When it finishes transmission, the next host is polled by the central controller. An enhancement limits the amount of time that a host can transmit so that everyone gets a fair chance to transmit.

A significant advantage of the centrally managed scheme is that it is possible for the hosts to inform the manager of the amount of traffic that they have. This is often measured in terms of how full their local queues are, but it can also be done on the basis of traffic priority, or rate of traffic arrival, for example. This means that the central controller has a global view of the state of the system, rather than the much more limited view a host would have, and can prioritise access by hosts accordingly. In the example in Figure 5.5, host C's queue is nearly full, and the manager gives priority for it to transmit. Since the manager has knowledge of access to the channel, it can predict the requirements of the hosts to a certain extent, reducing the need for hosts to inform it of their exact state.

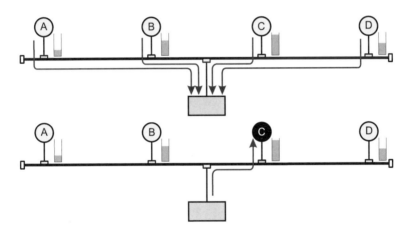

Figure 5.5   Priority-based polling scheme

One disadvantage of the polling scheme is that a host is polled whether or not it has something to transmit. If it has nothing to transmit, it will immediately signal this to the controller and the next host will be polled, but there is a significant overhead in this process if there are a large number of hosts who transmit infrequently. So-called hybrid schemes avoid this problem by using a contention-based scheme during set time periods with a polling scheme for most of the time.

In hybrid schemes, rather than polling all hosts, the controller keeps a polling list of hosts which are active, and only those hosts on that list are polled. Hosts use a contention-based scheme to inform the controller that they have something to transmit during periods set aside for this purpose. The controller will then put them on the polling list. When the host has finished transmitting they are again removed from the list.

5.3.2.2 Token-Based Schemes

Token-based schemes allow polling to take place in a distributed manner, so no central controller is required, although in practice one of the nodes has to generate the token initially and check that it is still flowing. In the scheme, a token is passed from node to node. Possession of the token gives permission to transmit data. In order to ensure fair access, the length of time a node can keep the token and transmit is limited. When the end of this time is reached, or when a node with the token has nothing to transmit, it passes the token on to the next node. This process continues in a cycle or logical ring until all nodes have had a chance to receive the token and transmit data, whereupon the cycle repeats.

There are many variations on the token theme, but a typical example of a token scheme is the IEEE 802.5 Token Ring MAC standard. In this system the token is formed from an empty information frame. If no data is being transmitted, the token or empty frame is passed from node to node. When a node has something to transmit, it waits for the token/frame, inserts the data it has to transmit and the destination address (as well as its own address as the source), and sends it on. Each node checks the information frame when it receives it to see if the destination address is its own. If not, it is passed on unchanged. When the destination address is reached, the destination node retrieves the information from the frame, and replaces the frame data with an acknowledgement, which it then puts back on the ring to be sent back to the source. On reaching the source, that node retrieves the acknowledgement (so it has confirmation of receipt), and then passes the empty frame back on the ring to the next node. This means that each node can only transmit once for each possession of the token, and if all nodes have something to transmit, the transmission will move round the ring from terminal to terminal in a round robin fashion.

As an example, consider the chain of events if A wants to send data to C in the system shown in Figure 5.6. A waits for free token. It then sets the token to be busy and forwards frame on ring to C. C receives frame and sets response bits in the tail of frame, putting it back on the ring to send to A. A receives frame, confirming the receipt of the data. The newly-freed token is returned to ring.

Another MAC, FDDI uses the same basic procedure although with a number of detailed differences. The main one is that with FDDI a node sends on a free token immediately after it has sent its data (i.e. before it receives an acknowledgement). In the larger networks that FDDI was designed for, waiting for an acknowledgement before freeing the token would be too inefficient. This leads to the possibility of a second (or even more) data transmission chasing the first or its acknowledgement round the ring, but only one transmission will be on any single part of the ring at any given time.

A significant advantage over the distributed contention-based schemes like CSMA/CD is that token-based systems can have different priorities while still operating in a distributed manner. This is carried out by allocating a priority field to the token and specifying that only services with a priority greater than the level given in the token can use it. To ensure that nodes transmit frames in priority order, an additional field to select the priority of the next frame to be transmitted is also needed. Again different token schemes implement the system in different ways, but the IEEE 802.5 system is fairly typical of the general procedure.

When a node has a frame to transmit, it calculates its priority. This will probably be related to the service the data is intended for, so speech, with its tighter delay constraints, would have a higher priority than a file transfer, for example. Each token has a current priority level and a reservation priority level. When a free token arrives at a node wishing to transmit with

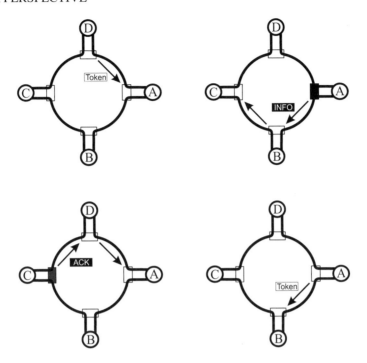

Figure 5.6 Token MAC scheme

a priority equal or lower than the priority of the traffic to be transmitted, the node uses the frame in the normal way. If the token priority is greater than the traffic priority, the token is passed on immediately.

To signal that it has priority information to send, nodes use the reservation field in the frame. If the reservation field has a lower priority than the data to be transmitted, the node inserts its priority level in the frame as its passes. When the token holder releases the token when it receives its acknowledgement, if the priority given in the reservation field is greater than the token priority, it raises the priority of the token to the reservation priority, ensuring that the token is passed round the ring until it reaches the high priority transmitter.

This priority mechanism means that transmission will operate frame by frame in a round robin manner for all users of the same priority level, and then move down to the next priority level when all those transmissions have finished, and so on. Any node that raises the priority level when it frees up a token in response to the reservation field is responsible for lowering it to its previous level next time it passes empty with the same high priority level. This would only happen when the high priority transmissions have finished, and avoids the situation of the network locking up with a high priority token but only low priority traffic awaiting transmission.

In the example in Figure 5.7, A is currently transmitting, while B has priority 3 data and C has priority 4 data. When the frame passes B, it is busy with A's traffic, so B fills the reservation field with 3. When it reaches C, C has a higher priority so replaces the reservation priority with 4. When A releases the token it therefore sets the priority to 4. The token passes B, but B sets the reservation to 3. C gets the token, and transmits its data. When it is finished, it releases a free token, but because it received the token with priority level 4, it leaves the priority at 4. The free token then circulates, but when it arrives empty at A it recognises that the high priority traffic has been sent and therefore reduces the priority level. Normally it would reduce this back to 0 where it found it, but because of the reservation for level 3, it issues a priority 3 token, remembering that it has now raised the token from priority 0 to priority 3, so it will reset the priority back to 0 next time it encounters an empty priority 3 token.

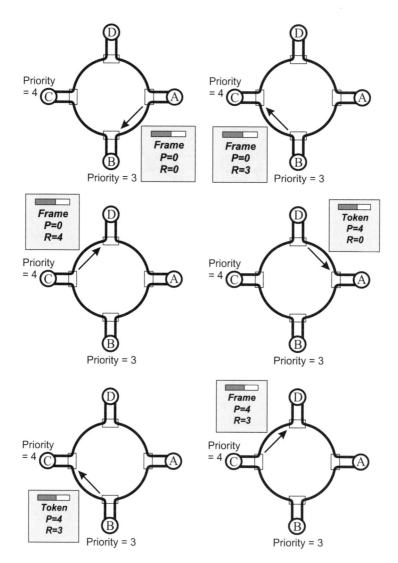

Figure 5.7   Token scheme with priorities

The priority scheme in FDDI is different because of the fact that its nodes transmit tokens immediately after sending their data, but the result is the same.

5.3.3   Comparison of Access Techniques

Of the various contention-based systems, CSMA/CD offers the highest throughput. Slotted ALOHA gives a higher throughput than ALOHA, and shares its advantage of not needing to know if another terminal is transmitting, so if such knowledge is impossible, Slotted ALOHA is used. If a contention-based scheme is used and reading the state of the medium is possible, CSMA/CD is the method of choice.

Of the reservation schemes, a central polling scheme is fairest and offers the best performance because the resource allocator has full knowledge of the state of the system. However, having this central control is often undesirable. Token-based systems offer the advantage of a distributed approach, but even with priorities, token systems only succeed in approximating the performance of a centrally based reservation scheme.

Comparing contention based schemes with reservation-based schemes, contention-based schemes are generally simpler and when the loading is light, offer faster access, as terminals do not have to wait for permission to transmit. However, it is not possible to assign different priorities to different traffic types in such systems, so they are poor for mixed traffic networks or networks with high priority traffic with low delay requirements. Access to the channel with the CSMA/CD is probabilistic which can result in variable and unbounded delay, and it is very unstable at loads above about 30%. Also, such schemes are comparatively inefficient and do not scale well, since as network sizes increase, so does the propagation delay.

For small networks like LANs, the distributed nature and low cost of CSMA/CD make it by far the most popular choice. The relative inefficiency is not important as over short distances the raw transmitted bit rate can be high to compensate. For larger networks, media costs are higher and become more significant than terminal costs, so a more efficient mechanism is preferable even if this results in more complex – and therefore expensive – terminals. Also, the requirement to give quality of service guarantees is more important. Therefore, larger networks use reservation-based schemes.

5.4   Channel Coding

If the channel introduces errors, the received messages will be equal to the transmitted messages with some errors added. For error detection, it is necessary to detect that these errors have been added, and for FEC, it is a further requirement to detect their location.

The code designer cannot control the errors, but it is still possible to define a code in such a way that the message can be recovered from the received codeword for most errors which will occur. In simple terms, if the messages are sufficiently 'different', they will still be recognisable as long as few errors occur. This 'difference' is introduced by adding redundant information to the message to form the codeword.

An example of adding to messages to make them more different is the phonic alphabet used spelling out words over a radio link. Many letters have very similar sounds – B, C and D for example – which can be confused when spoken over a noisy link. A phonic alphabet uses codewords for each letter to make them easier to distinguish. Table 5.1 gives the alphabet used by the British police. In addition, numbers are spoken with exaggerated syllables and with slight changes – nine to niner and five to fife – to make them easier to make out.

When designing a code to correct errors, the redundancy added in the form of additional symbols (bits in a binary code) have to be added with care to be effective. In particular, the redundancy should make similar codewords more different.

Table 5.1 Phonic alphabet

Letter (message)		Phonic 'codeword'	Letter (message)		Phonic 'codeword'
A	a	Alpha	N	en	November
B	bee	Bravo	O	o	Oscar
C	see	Charlie	P	pee	Papa
D	dee	Delta	Q	cue	Quebec
E	e	Echo	R	ar	Romeo
F	ef	Foxtrot	S	es	Sierra
G	gee	Golf	T	tee	Tango
H	aitch	Hotel	U	u	Uniform
I	I	India	V	vee	Victor
J	jay	Juliet	W	double-you	Whiskey
K	kay	Kilo	X	ex	Xray
L	el	Lima	Y	wye	Yankee
M	em	Mike	Z	zed	Zulu

The difference between codewords can be represented by the distance between them. Error correction can be undertaken by looking at the distance from the received codeword to all possible valid codewords, and choosing the nearest. If this is done, errors can always be corrected if they corrupt codewords by less than half the distance between the nearest two codewords, or they can be detected if they corrupt codewords by less than the smallest distance between two codewords.

Correction and detection can be carried out together. If the minimum distance between any pair of codewords is d_{min}, it is possible to correct errors up to d_c and detect errors up to d_d, as long as $d_{min} \geq d_e + d_c$, since if a received message lies outside the circles, it is possible to recognise that there is an error, although it cannot be corrected because it is not within d_c of any codeword. By varying the size of the circles, error correction capability can be traded off against error detection. Note that for a fixed d_{min}, increasing d_c reduces d_d, and one issue to be recognised is the possibility of incorrectly 'correcting' a received message. If a message was so badly corrupted that it was moved more than d_d, and so fell within d_c of another codeword, it would be decoded incorrectly.

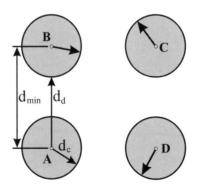

Figure 5.8 Distance between codewords

Mathematically, distance corresponds to a metric. The simplest and most common metric for binary signals is Hamming distance and is simply the number of coefficients where the binary bit streams differ. In other words, the Hamming distance between two bit streams a and b equals the weight of $a - b$, where the weight of a bit stream is the number of non-zero components it has.

Consider three codes using symbols made up of 3 binary digits, as shown in Figure 5.9. The first code uses all possible combinations of the three bits and so has a minimum distance

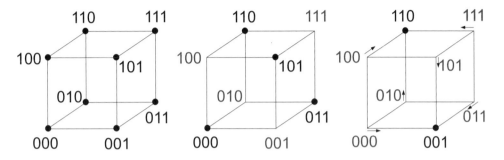

Figure 5.9   Three different codes with three binary digits

of 1. It cannot correct or detect any errors because any error would form a new codeword. Code 2 has a minimum distance of two, and was constructed by deleting all codewords with a distance of one from another codeword. This code can detect a single error, although not correct it (010 could be 000, 011 or even 110). Two errors would not be detected. Code 2 is actually an even parity check code, and will detect three errors as well. Code 3 has distance 3. It will correct one error or detect 2 (but not both). If it were used for correction, 010 would be corrected to 000, as would 100 and 001 as shown by the small arrows.

The existence of a distance between codewords does not necessarily imply that the message can easily be found from the received codeword, even if fewer errors occur than the code can theoretically correct. In the worst case, it may be necessary to compare the received bit stream with all possible codewords in order to see to which it is nearest. If the block length is large, there are many possible codewords, and so this can be a very time-consuming process. A choice of coding function must therefore be made so that there is a simple method of decoding.

For example, consider a relatively small (143,120) cyclic product code. This type of block code can be decoded using its coding rules in 264 simple steps (shifts). However, the code has 2^{120} codewords so an exhaustive search would require in the order of 1,000,000,000,000,000,000,000,000,000,000,000,000 comparisons.

There are several different methods of designing codes so that encoding and decoding can be carried out practically. If coding is a linear function with respect to the information message, we have a simple method of encoding and testing if received words are codewords. Repetition codes are very simple codes but are not very efficient. Hamming codes are a special class of single error correcting codes, which are simple to construct and use. Other common codes are cyclic codes and one type of these is Reed Solomon codes, which are complex but are now coming in to more general use (in CDs, for example). Convolutional codes operate on the data in a sequential manner, and while they are not as efficient as Reed Solomon codes, they are flexible and easy to design and use.

5.4.1   Probability of Error

Consider information transmitted through a binary symmetric channel with transition probability a. A binary symmetric channel has no memory and the probability of a transition (i.e. of an error) is the same for a 0 as it is for a 1. The probability that n bits are received without an error is $(1 - a)^n$. The probability that one error occurs is the probability than a given bit is in error with all the other bits correct (which is $a(1-a)^{n-1}$) times n, since there are n possibilities for the specific bit which is in error. More generally, the probability that e bits are in error from the n bits is

$$\binom{n}{e} = \frac{n!}{e!(n-e)!} = \frac{n(n-1)\cdots(n-e+1)}{i!}$$

More often we are interested in the probability that more than a given number of errors have occurred. This can be found easily from $P(\text{more than } e \text{ errors}) = 1 - P(e \text{ or fewer errors})$, which is given by

$$1 - \sum_{i=0}^{e} \binom{n}{i} (1-a)^{n-i} a^i$$

5.4.2   Constructing Error Correcting Codes

Unless otherwise stated we will assume binary codes. Error correcting codes can be constructed from other symbols but we will not consider any such codes in detail.

5.4.2.1   Repetition Codes

A very simple method of coding is to repeat the message $2d + 1$ times, and decode the received codeworda based on the majority of the received symbols. The distance between code words will be $2d + 1$, and this means that up to d errors can be corrected. This code is optimal for $d = 1$ and a message length of 1, i.e., a one bit error correcting code with codewords 000 and 111. However, when more binary symbols have to be transmitted, the performance of the code in terms of its rate for a given error correcting capability is very poor. Note that the definition of rate, R, for an error correcting code is (average number of source symbols)/(average number of codeword symbols). There is no concept of efficiency, but redundancy is defined as $1 - R$, (cf. source coding; the rate is considered equivalent to the efficiency and the source is assumed to have maximum entropy – one bit per binary symbol).

A binary repetition code capable of correcting 2 bit errors in six bits would need $d = 2$ and have a rate of 0.2. A convolutional code of type (2, 1, 3), detailed later, has the same error correcting capability but a rate of 0.5, and more complex block codes do better still.

5.4.2.2   Single Parity Check Code

One of the most common and simple form of codes is not actually an error correcting code since it can only detect a single error. A single parity check code is a binary code where one additional bit is added to each message to ensure that there is an even number of 1s. This means that a 0 is added to messages with an even number of ones and a 1 is added to messages with an odd number of 1s. Any single errored bit will leave an odd number of 1s so single errors can be detected, but two errors (or any even number of errors) will result in an even number of 1s and will go unnoticed. The code therefore detects one error and is not capable of any correction. Since all messages are different and all have an even number of 1s, the minimum distance is 2, which is in agreement with the correction ability.

Mathematically, if the message bits are $m_1, \ldots, m_n$, the parity bit is simply the modulo 2 sum of these bits $p = m_1 + \ldots + m_n$, so the parity check is $m_1 + \ldots + m_n + p = 0$.

5.4.2.3 Hamming code

A more scientific method of defining a code would be to consider that in order to resolve the error, the codeword needs to deliver enough information to contain the message and the error. Consider the need to correct a single error. If we look again at the three-bit codes in Figure 5.9, a single bit error could occur in any of these bits. Furthermore, there may not be any error at all. Therefore there are $3 + 1 = 4$ possible error messages. It takes two bits of information ($\log_2 4$) in order to encode this. The number of parity check bits – redundancy – added to the code must equal at least 2. There are three bits in total ($n=3$). Therefore, k, the number of information bits, equals 1.

In fact, code 3 in Figure 5.9 is a one bit error correcting code, with $k = 1$, and two codewords, 000 and 111.

We could extend our hypothesis as follows. If we have n bits in the codeword, we would need $\log_2(n + 1)$ parity check bits, leaving $k = n - \log_2(n + 1)$ information bits. It makes sense to make $n = 2^q - 1$ to make the most efficient use of the parity bits, so we could hypothesise that there are codes with parameters $n = 2^q - 1$, $k = n - q$ (i.e. (3, 1), (7, 4), (15, 11), etc.). It turns out that such codes do exist, and are called Hamming codes after their discoverer. 'Code 3' is a trivial example. The first interesting example is the (7, 4) code, which can correct one error in 7 bits, and has a rate of 57.1%.

The problem still exists as to how to construct these codes. A simple parity check can detect an error but not correct it. However, we have three parity bits. If we take different combinations of bits in order to form parities, when different combinations fail, we can find the error. With 3 parity bits we have 8 combinations. Let the 000 combination, i.e., no parity failures, correspond to no errors. Let each parity bit check itself and three of the four message bits. Then three of the message bits are checked by two parities, and one by three parity check bits, while each parity check bit is only checked by itself. A failure of one parity check will show an error in the parity check bit. A failure in two will show a failure in the corresponding message bit appearing in the two parity checks, and a failure of all three parity check bits would indicate an error in the bit checked by all three parity check bits.

5.4.3 Linear Codes

While the repetition code was easy to design, as we construct more efficient codes, things become more complicated and we are forced to design codes more systematically. A linear code is one where each codeword is formed from a linear combination of the message vector.

A two-dimensional code can be formed, defined such that a message i is encoded into a two-dimensional codeword $(i, 2i)$. This is a linear code, since every linear combination of codewords forms another codeword. The best codes are formed by those functions which cause a wide spread of codewords with a large distance between them. This example forms a one error correcting code, if errors in this case can change one dimension by one unit. Let

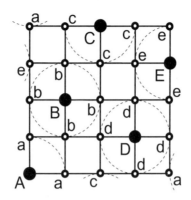

Figure 5.10   Simple linear codes with two dimensions taken modulo 5

the five codewords be lettered A, B, C, D, and E, in Figure 5.10. Each of the remaining grid points is at most one horizontal or vertical position away from a codeword. This can be seen by labelling each non-codeword with the small letter of the codeword to which it is nearest. If we receive any of the small letters we assume that the corresponding capital letter was sent.

Linear codes are often defined by a generator matrix made up of code words. k of these codewords are independent, so the generator matrix G has k rows and n columns. All the codewords can be formed by adding rows, so the codeword, c, is formed from the message, m, by $c = mG$. The structure of G is such that when it is put in standard echelon form it has the structure $[I|P]$. There is a corresponding matrix $H = [P^\top|I]$ called the parity check matrix. $GH^\top = 0$. H is used to decode the code. If there are no errors, $cH^\top = mGH^\top = m0 = 0$. If an error, e, occurs, we receive $c' = c+e$. $c'H^\top = (c+e)H^\top = cH^\top + eH^\top = 0 + eH^\top = eH^\top$. This value, called the syndrome, depends only on the error and not on the codeword transmitted. This is a key property of linear codes.

The syndrome is used to correct the errors. A syndrome of zero indicates that a codeword was received. This means either that no error occurred, or that the error is a codeword, since then the linear property of the code is such that $c' = c + e$ will also be a codeword. Any error having the same form of the codeword would have a weight equal or exceeding the minimum distance of the codeword, so could not be corrected anyway.

Linear codes are much easier to implement because, rather than having to record details of every codeword, we can simply record details of the linear function, and multiply the message by that. However, error correction is not, in general, simple. Some linear codes do have defined methods to detect the nearest codeword in the presence of errors, but in general the better the code (in terms of the number of errors it corrects and the additional bits added), the more complicated the algorithms required to decode the code.

5.4.3.1   Matrix Form of Hamming Codes

Recall the definition of Hamming codes. The parity check equations formed a linear combination on the codewords, so Hamming codes are linear, and can be expressed in matrix form. In fact, the parity check matrix of the code we formed above is as follows:

$$p_1 = m_1 + m_2 + m_4 \Rightarrow m_1 + m_2 + m_4 + p_1 = 0$$
$$p_2 = m_1 + m_3 + m_4 \Rightarrow m_1 + m_3 + m_4 + p_2 = 0 \qquad H = \begin{pmatrix} 1 & 1 & 0 & 1 & 1 & 0 & 0 \\ 1 & 0 & 1 & 1 & 0 & 1 & 0 \\ 0 & 1 & 1 & 1 & 0 & 0 & 1 \end{pmatrix}$$
$$p_3 = m_2 + m_3 + m_4 \Rightarrow m_2 + m_3 + m_4 + p_3 = 0$$

However, moving up to the (15,11) code is difficult, because of the lack of a simple design rule. However, note that in row echelon form, each row differs by at least two (due to the I part of the matrix). If each row in the P part of the matrix is different, and contains at least 2 bits (to ensure the minimum distance is 3; any row with fewer bits would be fewer than 3 bits from the all 0 codeword), the result will be a matrix representing a combination of checks which satisfy all our requirements. Therefore, to construct a Hamming code, we simply fill P with every combination of bit sequences with 2 or more 1s, in any order. (Different orders will give different, but equivalent, Hamming codes.)

Decoding Hamming codes is very simple using the H matrix as they are single error correcting. Any codeword postmultiplied by the transpose of the H matrix gives a zero vector. If there is a single error, the received codeword c' can be considered to be equal to a single bit vector e and the codeword c. $c'H^\top = (c + e)H^\top = cH^\top + eH^\top = 0 + eH^\top$. However, since e is a vector with only a single 1, $eH^\top$ is just a column in the H matrix, the position of which corresponds to the error. (This argument applies to all single bit error correction by a linear code, not just to Hamming codes.)

5.4.3.2   Cyclic Codes

Special types of linear codes, called cyclic codes, have a well-defined structure and are simple to implement. A cyclic code is a code where in addition to the property that every linear combination of codewords forms another codeword, every cyclic shift of the codeword forms another codeword. For example, in the binary case, if 01001011 is a code word, so is 10010110, 00101101, and so on. Such codes can be therefore be constructed from shifts and linear combinations of a single codeword, called a generator. If properly designed, the generator ensures that the codewords are evenly distributed throughout the code space and the code has good distance properties. It is also easy to check if a received codeword has any errors, as all codewords are divisible by the generator. For this reason, special cyclic codes called cyclic redundancy checks (CRC) are used for error detection and not error correction.

5.4.3.3   Linear Code Example: Extended Hamming Code

A (7,4) single error correcting Hamming code has the following generator matrix:

$$\begin{pmatrix} 1 & 0 & 0 & 0 & 1 & 1 & 0 \\ 0 & 1 & 0 & 0 & 1 & 0 & 1 \\ 0 & 0 & 1 & 0 & 0 & 1 & 1 \\ 0 & 0 & 0 & 1 & 1 & 1 & 1 \end{pmatrix}$$

We can add an overall parity check to this code to form a (8,4) code which has a minimum distance of 4.[1] This is done by coding the output of the Hamming code with an overall

---

[1] We can add an overall parity check to any code which does not already have one, and the distance will be increased by 1. If the code already has an overall parity, all the codewords will already have an even weight, so the parity bit would always be 0, and the distance and error correcting capability would be unchanged.

parity check. The Hamming code output is $c = mG$, where G is the generator matrix of the Hamming code.

$$\begin{pmatrix} c_1 & c_2 & c_3 & c_4 & c_5 & c_6 & c_7 \end{pmatrix} \begin{pmatrix} 1 & 0 & 0 & 0 & 0 & 0 & 0 & 1 \\ 0 & 1 & 0 & 0 & 0 & 0 & 0 & 1 \\ 0 & 0 & 1 & 0 & 0 & 0 & 0 & 1 \\ 0 & 0 & 0 & 1 & 0 & 0 & 0 & 1 \\ 0 & 0 & 0 & 0 & 1 & 0 & 0 & 1 \\ 0 & 0 & 0 & 0 & 0 & 1 & 0 & 1 \\ 0 & 0 & 0 & 0 & 0 & 0 & 1 & 1 \end{pmatrix} =$$

$$\begin{pmatrix} m_1 & m_2 & m_3 & m_4 \end{pmatrix} \begin{pmatrix} 1 & 0 & 0 & 0 & 1 & 1 & 0 & 1 \\ 0 & 1 & 0 & 0 & 1 & 0 & 1 & 1 \\ 0 & 0 & 1 & 0 & 0 & 1 & 1 & 1 \\ 0 & 0 & 0 & 1 & 1 & 1 & 1 & 1 \end{pmatrix} \begin{pmatrix} 1 & 0 & 0 & 0 & 0 & 0 & 0 & 1 \\ 0 & 1 & 0 & 0 & 0 & 0 & 0 & 1 \\ 0 & 0 & 1 & 0 & 0 & 0 & 0 & 1 \\ 0 & 0 & 0 & 1 & 0 & 0 & 0 & 1 \\ 0 & 0 & 0 & 0 & 1 & 0 & 0 & 1 \\ 0 & 0 & 0 & 0 & 0 & 1 & 0 & 1 \\ 0 & 0 & 0 & 0 & 0 & 0 & 1 & 1 \end{pmatrix} =$$

$$\begin{pmatrix} m_1 & m_2 & m_3 & m_4 \end{pmatrix} \begin{pmatrix} 1 & 0 & 0 & 0 & 1 & 1 & 0 & 1 \\ 0 & 1 & 0 & 0 & 1 & 0 & 1 & 1 \\ 0 & 0 & 1 & 0 & 0 & 1 & 1 & 1 \\ 0 & 0 & 0 & 1 & 1 & 1 & 1 & 1 \end{pmatrix}$$

The generator of the overall (8,4) code, called an extended Hamming code, is therefore

$$\begin{pmatrix} 1 & 0 & 0 & 0 & 1 & 1 & 0 & 1 \\ 0 & 1 & 0 & 0 & 1 & 0 & 1 & 1 \\ 0 & 0 & 1 & 0 & 0 & 1 & 1 & 1 \\ 0 & 0 & 0 & 1 & 1 & 1 & 1 & 1 \end{pmatrix}$$

The codewords are

00000000	00100111	00011110	00111001
10001101	10101010	10010011	10110100
01001011	01101100	01010101	01110010
11000110	11100001	11011000	11111111

The codewords can either be generated directly from the generator matrix, or by listing the codewords of the Hamming code, and adding a final bit to give an even number of 1s.

If there is no error the syndrome is 0 (strictly, 00000000). The syndromes for all single errors are

Error	Syndrome
10000000	1101
01000000	1011
00100000	0111
00010000	1110
00001000	1000
00000100	0100
00000010	0010
00000001	0001

All the syndromes are distinct, so all single errors can be corrected.
There are 28 possible double errors.

Errors	Syndrome	Errors	Syndrome	Errors	Syndrome	Errors	Syndrome
11000000	0110	01100000	1100	00101000	1111	00010001	1111
10100000	1010	01010000	0101	00100100	0011	00001100	1100
10010000	0011	01001000	0011	00100010	0101	00001010	1010
10001000	0101	01000100	1111	00100001	0110	00001001	1001
10000100	1001	01000010	1001	00011000	0110	00000110	0110
10000010	1111	01000001	1010	00010100	1010	00000101	0101
10000001	1100	00110000	1001	00010010	1100	00000011	0011

While not all the syndromes are distinct (for example, 11000000, 00100001, and 00011000 produce the same syndromes), they are all different from the syndromes for single errors. This means that while we cannot correct two errors, we can detect them *in addition* to correcting single errors. This contrasts with the standard Hamming code, where we could *either* correct one error *or* detect two errors. For the extending Hamming code, we can correct one and detect two, *or* detect three.

5.4.4 Convolutional Codes

The codes discussed so far are block codes, where information is processed a block at a time. An alternative is to encode and decode sequentially. Here a number of output bits are generated each time an input bit is received. A single bit repetition code does this – it simply repeats the input bit several times. If the rate of repetition is 2, we would get the following:

Input   1   1   0   1
Output   11   11   00   11

This is actually a very poor code. It has a rate of 1/2, but only has a distance of 2, and so can only detect one error. Performance can be greatly improved by adding some memory to the coder and making the current output bits dependent not only on the current input but also on previous bits. Such codes are called convolutional codes. The number of bits taken into account – the memory of the code + 1 – is called its constraint length.

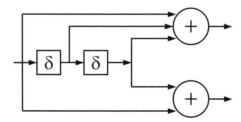

Figure 5.11   (2, 1, 3) convolutional code

A simple convolutional code is a (2, 1, 3) code (2 output bits for each 1 input bit with a constraint length of 3) such that the first bit is the binary sum of the current input, the input on the last bit, and the input two bits previously, and the second bit is the sum of the current bit and the bit two bits ago. A suitable circuit to generate this code is shown in Figure 5.11. The code has a distance of 5, and so it can correct two errors. It can correct more errors than this if the errors are sufficiently widely spaced (i.e. further apart than the code's memory).

Convolutional codes are simple to encode and decode, but if they do make a mistake, then a large number of output bits will be corrupted (since the decoder will attempt to correct the bit stream and add additional errors in the process). For this reason, some form of additional error detection is desirable.

The memory in the shift registers of a convolutional coder gives the system a state. For the (2,1,3) convolutional code there are $2^{L-1} = 4$ states. A state transition diagram has a node for each possible state of the system, connected by lines depicting the possible changes of state the system can undergo. These lines are labelled with the inputs which would cause this change of state, and the output that is produced. The state transition diagram for the (2, 1, 3) code is shown in Figure 5.12. For example, if the coder is in state 00, it can remain in state 00 by receiving a 0, whereupon 00 will be output, or it can move to state 10 if a 1 is received, outputting 11 as it does so.

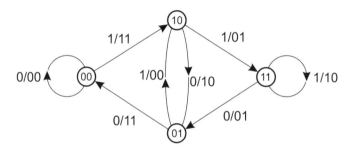

Figure 5.12   (2, 1, 3) code state transition diagram

While a state transition diagram is useful for showing possible movements between states of a system, for a convolutional code it is usually more important to consider a sequence of state changes and corresponding outputs in response to different inputs, and for this purpose a code trellis is useful. This is similar to a state transition diagram but with the existing states listed on the left and the next states on the right, with state transitions drawn in between (see Figure 5.13).

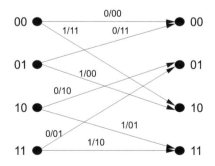

Figure 5.13   Trellis diagram for the (2, 1, 3) code

The example in Figure 5.14 shows the states and output of the (2, 1, 3) code when the binary stream to be encoded is 101001011100. The state transitions are denoted by the thick line.

Step Number	1	2	3	4	5	6	7	8	9	10	11	12
Input Bit	1	0	1	0	0	1	0	1	1	1	0	0
Current State	00	10	01	10	01	00	10	01	10	11	11	01
Next State	10	01	10	01	00	10	01	10	11	11	01	10
Output Bits	11	10	00	10	11	11	10	00	01	10	01	11

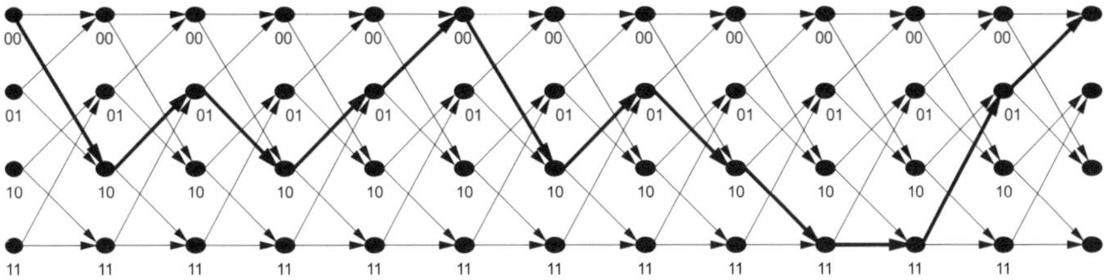

Figure 5.14   Example of encoding using the (2, 1, 3) code

There are several different methods of decoding convolutional codes and, as would be expected, there is a trade-off between decoding complexity and performance. A relatively simple method, called sequential decoding, is to trace through the encoding trellis based on the received bits, taking the path at each node with results in the lowest number of discrepancies between the received bits and the bits that would have been generated. This number is termed the branch metric. If two paths have the same metric, a random choice is made between them and decoding continues. If the metric exceeds a threshold based on the expected number of errors, the decoder decides that it must have made a mistake on one of its random choices, deletes that path, and tries again.

This approach can be extended to give maximum likelihood decoding (i.e. choosing the transmitted sequence which would produce the smallest number of differences between its codeword and the received sequence). If every possible path from each node is followed, then all possibilities are taken into account. However, if two or more paths reach the same node, they will all continue in exactly the same way, so it is possible to delete all arriving paths except the one with the lowest metric. This means that the number of surviving paths is kept to, at most, the number of states, and so the complexity is reasonable. This algorithm is the Viterbi algorithm, and while it is more complex than the sequential algorithm, it can be implemented quite easily using modern techniques.

One problem with convolutional codes is that they have a rather poor rate, usually less than 50%. The rate can be improved by 'puncturing' the code. Puncturing is a process of removing some of the additional error correcting bits. This reduces the ability of the code to correct errors, but the process enables us to find a compromise between the number of bits we add, and therefore the power of the code to correct errors, and the transmission requirements in terms of number of bits.

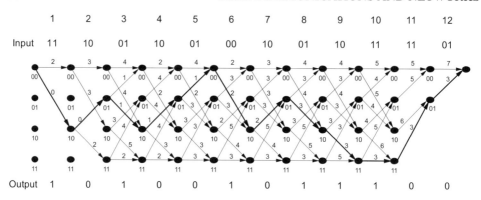

Figure 5.15 Example of decoding using the (2, 1, 3) code

5.4.5 Interleaving

A binary symmetric channel introduces errors at random. This is a reasonable estimate of the performance of some channels (in particular channels subject to *white noise*), but many channels have memory and errors are produced in bursts (radio channels in particular have this property). Most error correcting codes (including Hamming codes and convolutional codes) are most effective against errors which occur randomly. This is because if the errors occur in groups there are too many to correct in a particular interval even though on average the error rate is within the correcting capability of the code. To counter this interleaving is used to spread the bits so that adjacent transmitted bits are spread out before coding (see Figure 5.16), limiting the number occurring together to within the correcting ability of the code. If the errors, when spread, exceed the capability of the error correcting code, the code will introduce more errors as it unsuccessfully attempts to correct them. This means that if the average error rate is more than the code can correct, interleaving will reduce performance, since it spreads the errors which can not be corrected into previously error-free codewords.

5.4.6 Combining Codes

Different types of coding can be combined. This can often be useful in order to combine the advantages of the different constituent codes.

5.4.6.1 Product Codes

One of the simplest ways of combining codes is to form a product code. The symbols to be encoded are arranged in an array, as shown in Figure 5.17, and then one type of block code is applied to the rows and another to the columns. It is possible to combine more than two codes, but a product of two codes is the most common arrangement.

The constituent codes need not be different, and a simple product code is the product of two single parity check (SPC) codes, which are sometimes referred to as Gilbert codes. SPC codes cannot correct an error on their own, but in a product code a burst of errors can be identified by cross-referencing the parity checks (see Figure 5.18). If the number of rows is m_1 and the number of columns is m_2, this code has $n = m_1 m_2$, $k = (m_1 - 1)(m_2 - 1)$ and is capable of correcting a single burst of errors of length up to $m_1 - 1$.

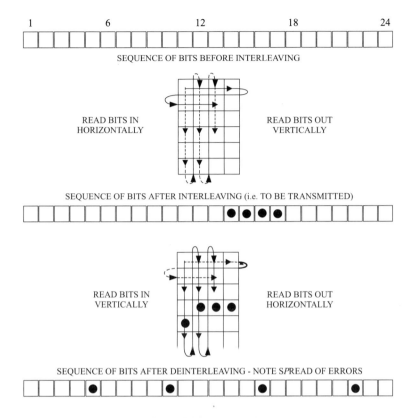

Figure 5.16   Interleaving

Message Symbols	Row Code Parity Checks
Column Code Parity Checks	Checks on Checks

Figure 5.17   Product code

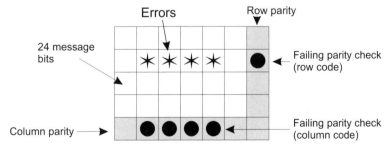

Figure 5.18   Product of SPC codes

5.4.6.2   Concatenated Codes

Another method of combining codes is to apply two or more codes to the data separately in a process referred to as concatenation. Concatenation can take two forms. The most common is serial concatenation, where one code is applied to the data to be transmitted, and then the second code is applied to the output, message and parity, of the first code. Often interleaving is used prior to the application of the second code, as shown in Figure 5.19.

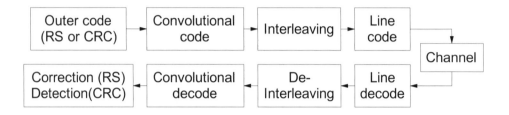

Figure 5.19   Serial concatenation

For example, convolutional codes are some of the most commonly used channel codes. Interleaving and error detection are also used to improve performance and detect any errors the convolutional code has not dealt with. Reed Solomon codes, efficient linear block codes which are good at correcting bursts of errors, are often used for additional error correction. Encoding is performed in the sequence block error correction or error detection, followed by random error correction, followed by interleaving, followed by transmission, with the sequence being reversed on reception (see Figure 5.19). Interleaving is performed at the lowest level, closest to the channel, so that any bursts of errors caused by the channel are spread so that the random error correcting code, usually a convolutional code, can correct them. If the convolutional code is overwhelmed with errors, it is likely to make a mistake in correcting, generating a block of errors, which the outer code can correct if it is a block error correcting code. More commonly a cyclic redundancy check is used, which simply detects whether the received data is valid.

The second method of concatenation is parallel concatenation, shown in Figure 5.20. Here the different forms of coding are both performed at the same time on the data. This means that the parity symbols of one code and not encoded by the other. Interleaving is used to vary the relationship between the message symbols and make the combination more effective. Depending on this interleaving, a parallel concatenation of two codes is similar to a product code, except that the parity checks on the parity checks are not present.

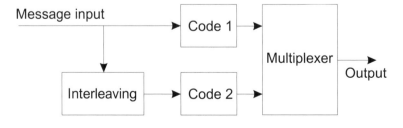

Figure 5.20   Parallel concatenation

5.4.6.3 Turbo Codes

The parallel concatenation principle has a significant bonus. Since we have two codes operating on the data, we can use information one code provides us about the position of errors to assist the other code in its error correction. We can apply this process iteratively, so that we can make several passes exchanging information between the different codes. Consider two codes operating on rows and columns each capable of correcting one error. The error pattern in Figure 5.21 would defeat this code, but a first pass would allow two of the errors to be corrected by the row code (Figure 5.21 (b)), which would then allow the column code to clean up the data and remove the remaining errors (Figure 5.21 (c)).

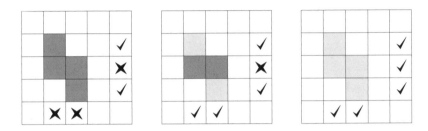

Figure 5.21   Iterative decoding to correct errors

This procedure, was invented by Berrou, Glavieux and Thitimajshima for their 'Parallel-concatenated recursive-systematic convolutional codes', which they also gave the shorthand name 'Turbo Codes'. It was first applied to systematic convolutional codes, which are convolutional codes with the encoder modified so that its output is formed of separate message and parity bits, like the code in Question 1 of Section 5.11.2, but the turbo principle can be applied to other types of code as well. In fact, it can be applied directly to product codes simply by changing the decoder. Product codes with an iterative decoder are called Turbo-product Codes.

The main interest in Turbo Codes comes from the fact that they have very high performance, and approach the theoretical limits of transmission capacity given by the Shannon bound (see Section 6.2.4). The disadvantage of Turbo Codes is that to approach the limit closely requires a large interleaver, which therefore introduces a large delay, although this is less of an issue with Turbo-product Codes. Classical Turbo Codes are not very suitable for speech transmission, but modern data transmission applications are making increasing use of this new technology.

5.5 Comparison between Forward and Feedback Error Correction (ARQ)

Having examined ARQ, it is interesting to consider how it compares with the feedback error correction strategy discussed in Section 4.11.2.

A disadvantage of FEC is that the additional redundancy must be dimensioned so that the worst case of expected errors can be corrected, and if fewer errors occur, this redundancy is wasted. The result is a lower channel capacity than would otherwise be obtainable.

When ARQ is used, if an error occurs, the receiver requests that the transmitter resends the information. ARQ schemes will only work if there is a reasonable chance that a block will be transmitted without error. If a block is likely to be in error, this will also be the case when it is retransmitted. ARQ schemes work best where there are occasional severe errors which occur infrequently but would require too much redundancy if FEC was used.

The amount of redundancy used for FEC can be compared with that required for ARQ. In the FEC case, the redundancy is fixed at the number of error correcting bits in the code. In an ARQ scheme, there is also a fixed redundant element – the error detection bits – but these will be fewer than in the FEC case. However, in the case of ARQ, errored blocks are also effectively redundant, so if blocks have to be retransmitted too frequently the overall throughput will be lower than the FEC case. This is shown graphically in Figure 5.22, with the redundant information shaded.

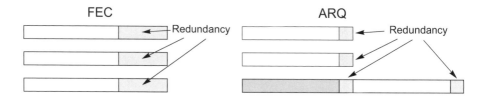

Figure 5.22   Comparison of redundancy with FEC and ARQ

ARQ systems tend to be less complex than FEC schemes, as error detection is simpler than error correction. However, ARQ schemes require a return path to the transmitter, which must have a low error rate. Also, while FEC schemes have a fixed delay, ARQ introduces a variable delay depending on the number of retransmission which are required. This is not ideal for constant delay services like speech.

It is possible to combine FEC and ARQ. FEC is the most efficient strategy when the error rate is reasonably constant, while ARQ is good for variable error rates. If the error characteristics of the channel are such that there is a constant background error probability, along with some bursts of higher numbers of errors, a FEC scheme can be used within an ARQ scheme with the FEC designed to clean up the background errors conforming to the constant error rate. Should a burst of errors occur, the FEC will fail, and the ARQ scheme will then operate to resend the information.

A development of such a system is a so-called 'hybrid' ARQ scheme. Again FEC is employed, but if it fails, instead of resending the block, additional error correcting information is sent which is combined with the data already sent in order to try to correct the errors. Hybrid ARQ schemes require complex error correcting coding schemes, but are very efficient for services which can tolerate a variable delay.

5.6   Local Area Networks

Local Area Networks are, essentially, relatively high-speed data networks which interconnect computing devices within a small geographical or local area. They are generally owned by a single organisation, and are independent of public networks.

LANs are private networks in the sense that they serve a single organisation within a limited environment. The costs associated with a LAN are the capital and installation costs which are met by the users (owners). The local environment usually refers to a site with a maximum span of a few kilometres, housing a relatively small number of buildings.

There are a number of requirements on a LAN design:

- It must be capable of handling data traffic, as this is the main traffic type within a local area.
- It must provide high transfer rates. Most communication requirements are within the local area, so local area networks will have higher data rates than wide area networks. Speeds at least in the order of 10Mbit/s are required, rising to hundreds of megabits per second.
- It must be easy to manage, since the operators of the network are the users themselves, not a specialist communications company. This generally requires some form of distributed control, and a common transmission medium with shared access.
- It should be low cost, both for installation and in particular for maintenance. This requires the use of a common standard giving device, control, and manufacturer independence, as well as a low cost media.

Set against these requirements is the fact that the network has to operate over a relatively small area, and since it is independent of public operators, there is more freedom to choose a standard.

5.6.1 LAN Configuration

Early LANs were usually bus-based systems since this allowed terminals to be added anywhere along the bus for easy management and avoided the additional complexity of switching nodes. However, a moderately expensive transmission medium, in the form of coaxial cabling, is required in this case. With advances in signal processing and reduced component costs, low costs hubs allow the use of cheaper unshielded twisted pair wiring at higher speeds, so a physical star is now a popular configuration.

Optical fibre is unsuitable for a bus-based system, since it requires a point-to-point link. Optical fibre-based LANs therefore have a ring or star configuration.

Whether physically a bus or a star, LANs use a common transmission medium which while requiring medium access control does significantly reduce management complexity to the point where some systems can be considered plug in and go.

LANs are packet-based networks, as packet switching is more suitable than circuit switching for handling bursty and unpredictable data traffic from a diverse range of sources.

5.6.2 IEEE 802 LANs

The most popular LAN standards are the IEEE 802 series. These standards (named after each sub-committee) define a family of LANs differentiated by their medium access control scheme. Figure 5.23 shows a simplified relationship between various IEEE standards and the OSI model. There are currently more than twenty IEEE standards.

Standard IEEE 802.1 covers both layers 1 and 2. This standard provides the framework for higher layer issues and essentially puts the other standards into context. Standards 802.3, 802.4 and 802.5 define the MAC and physical layer specifications for three different LAN types: CSMA/CD, Token Bus and Token Ring. The MAC elements define how the physical medium is accessed and provides a subset of data-link functions.

The upper-layer data-link functions are defined in IEEE 802.2. This standard provides a consistent interface between any LAN MAC and higher layer protocols. Put another way, the IEEE 802.2 Logical Link Control (LLC) provides a uniform protocol interface between higher layers and the actual underlying network, and makes the MAC and the LAN implementation transparent to higher layers, i.e. the application.

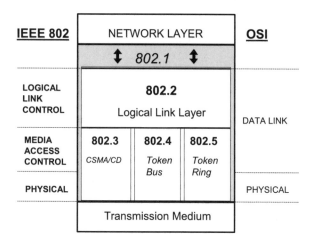

Figure 5.23   IEEE 802 protocol stack

The LLC makes a broadcast network appear to the network layer as a point-to-point link. It provides three classes of service:

- **Unacknowledged connectionless service** is a datagram service that supports only sending and receiving frames. Its simplicity makes it the easiest to implement and useful in situations where higher layers provide end-to-end error and flow control
- **Connection orientated service** provides a virtual circuit style connection between service access points. Allows user to request setup and termination of a logical connection. Also provides error and flow control. Useful in applications where device is relatively simple, with few if any, upper layer protocols e.g. terminal.
- **Acknowledged connectionless service** provides a mechanism by which delivery of a frame is acknowledged without the necessity to set up a connection. This is useful in real time applications where signals require acknowledgement.

5.6.3   Device Addressing

The IEEE standards defines a 48 bit MAC address which is divided into 4 parts, as shown in Figure 5.24. The first two bits indicate whether the frame is uni-cast (0) or multicast (1) and is universally (0) or locally managed (1). The third field indicates the IEEE defined part of the address which manufacturers register with the IEEE to obtain, while the last 3 octets are selected by the device manufacturer. The address is globally unique to the device and is usually hardwired.

5.6.3.1   Address Resolution Protocol

The link layer address (for example, defined by IEEE 802) is the address a host's network card will recognise. This is called the network point of attachment (NPA) address. The higher layers will use a different address, usually an IP address, so the packet from the higher layers with that address must be encapsulated in a link layer packet with a link layer address (see Figure 5.25). There is therefore a need for a need for a mapping between the two address types. For IP, this is undertaken by the Address Resolution Protocol.

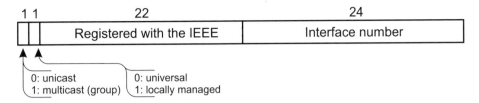

Figure 5.24   IEEE 802 MAC address

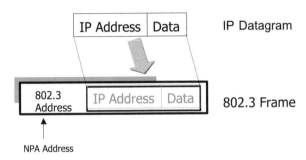

Figure 5.25   IP packet and its address encapsulated in an Ethernet packet

Hosts and gateways store IP/NPA address pairs for all hosts on the same network in an ARP table. ARP is used to inform stations of these address pairs and to provide 'translations' between IP and NPA addresses and vice versa. In the example in Figure 5.26, the gateway receives an IP datagram for a host whose NPA address it does not know. It therefore broadcasts a ARP request packet containing its own IP/NPA pair and the IP address of the node it is looking for. The host will recognise itself, and send back an ARP reply message with its IP/NPA pair. As well as knowing where to send that packet, the gateway will update its ARP table for future use. ARP table entries periodically expire to allow for changes in the network. A useful feature of ARP is that the ARP tables are built automatically and do not have to be manually set up.

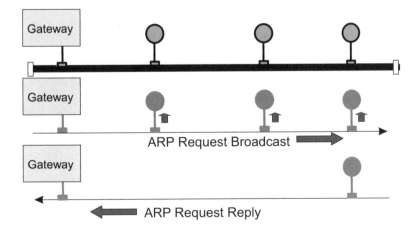

Figure 5.26   ARP operation

5.6.3.2 Reserve Address Resolution Protocol

The Reserve Address Resolution Protocol, as the name implies, performs the reverse mapping from hardware address to IP address. It is often very useful not to store IP addresses on hosts, but rather to have one centrally managed list. This allows addresses to be changed easily and copes with the problems of diskless workstations which cannot keep a permanent record of their IP address. The protocol essentially applies at boot up when the node broadcasts a message (to the server) to say 'does anyone know my IP address?' – the server receives this message and looks up its tables (or asks someone else) for the information and then passes it back to the node. Other stations on the network can also hear this dialogue and store the information locally in their ARP table.

RARP allows IP addresses to be shared among infrequently used hosts. However, since the hosts would not have permanent IP addresses, they would be difficult to contact, so would be unsuitable for running programs like web servers. On the other hand, such services would normally be hosted on nodes which are on permanently.

5.6.4 IEEE 802.3 (Ethernet) LAN

IEEE 802.3 describes the physical layer and MAC for a Carrier Sense Multiple Access with Collision Detection (CSMA/CD) LAN. This standard is based upon the Ethernet LAN developed by Xerox and often the names Ethernet and CSMA/CD are considered to be synonymous. While there are a great many similarities between Ethernet and CSMA/CD, they are not exactly the same.

CSMA/CD is a bus-based LAN (a logical bus, but not always a physical bus). Manchester line coding is used on a variety of different media, as follows:

- **10Base-5:** 10 Mbits/sec line rate using 'thick' coaxial cable. This is the original version and has a maximum segment length of 500m.
- **10Base-2:** 10 Mbits/sec line rate using 'thin' coaxial cable. Often referred to as Cheapernet and has a maximum range of 185m per segment.
- **1Base-5:** 1 Mbits/sec line rate using twisted-pair cable. A physical star topology is adopted, but logically acts as a bus.
- **10Base-T:** 10 Mbits/sec line rate using twisted pair. Again a physical star topology is used.
- **10Base-F:** 10 Mbits/sec line rate running over optical fibre. Star-based.
- **10Broad36:** Broadband version operating over coaxial cable.

Figure 5.27 shows a sample configuration of a CSMA/CD network. The network is logically represented by a single bus. If a node has information to transmit, the transceiver transmitter 'listens' for traffic on the link – carrier sensing. If there is no traffic on the bus, one frame of data is transmitted. If the bus is being used, i.e. busy, the transceiver waits until the link is idle and then transmits. As a broadcast network, all MAC frames are received by all nodes. A transceiver identifies a frame intended for that node by the address field of the received frame. Unwanted frames are discarded.

Figure 5.28 shows the IEEE 802.3 Frame Format. A variable length frame is used. The frame starts with 8 bytes of preamble, consisting of 7 bytes of 10101010 and a final one of 10101011, so '11' announces the start of frame. This is followed by the address of the destination and the address of the source, both in the 6 byte NPA format. The next two bytes give the length of the data, then comes the data itself, and finally a 4 byte frame check

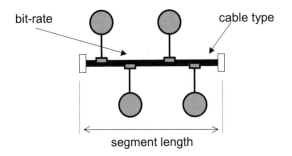

Figure 5.27   IEEE 802.3 bus structure

sequence. If the data is less than 46 bytes it is padded to 46 octets, so that the complete frame is long enough for the CSMA/CD access scheme.

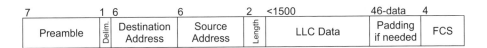

Figure 5.28   IEEE 802.3 frame format

The original Ethernet frame was slightly different from the IEEE 802.3 frame format. The definition specified an 8 byte preamble rather than 7 bytes and a start of frame delimiter, but the actual bits defined are the same. However, the length field was originally a type field defining the protocol, such as IP, in use. Compatibility is maintained by the fact that all the originally defined type codes used numbers greater than 1500, so a network adapter can distinguish an Ethernet frame from an IEEE 802.3 frame. If different protocols are in use in an IEEE 802.3 network, this is indicated at the start of the data field.

5.6.5   High Speed Ethernet

High Speed Ethernet refers to a range of network types based upon hub-based CSMA/CD access methods. Although the original CSMA/CD operation was limited by physical factors, hub-centric architectures do not have the same restrictions – the bus is a logical one rather than a physical one so does not have the same propagation issues. High speed access is achieved through a combination of improved transmission media and coding methods plus the very practical approach of increasing the number of wires.

There are three categories of higher speed Ethernet:

- IsoEthernet
- Fast Ethernet
- GigaBit Ethernet

5.6.5.1   IsoEthernet

Isochronous Ethernet is essentially an extension of 10Base-T standard which includes an additional 6.144 Mbits/sec of isochronous capacity for multi-media time sensitive traffic. It is defined in the IEEE 802.9 standard. IsoENET can support ninety-six 64 kbits/sec synchronous channels – in line with voice provision over the telephone network.

The 20 Mbits/sec clocked signal is carried over twisted pairs but the Manchester coding is replaced with 4B5B coding, which is used in FDDI. This is 80% efficient compared with Manchester coding's 50%.

The system is flexible and upgradeable. It requires an isoENET hub but users not requiring isochronous support can use their 10Base-T adapter cards. Only users requiring isochronous support need purchase an adapter card.

5.6.5.2   Fast Ethernet

Fast Ethernet is term that generally refers to the three 100 Mbits/sec physical layer specifications that form part of the IEEE 802.3μ addendum. The three techniques are differentiated by the coding techniques used and the transport media. All three are compatible with all other IEEE 802 standards.

The changes in media and coding technique are possible because of the use of a physical star rather than a physical bus. Manchester coding uses a clock rate that is twice the required data rate. 4B5T takes 4 binary digits and replaces these with 5 binary digits. The 'extra' bit is used for protection – essentially a block code with 80% efficiency. It carries 1.6 times more data than Manchester coding using the same clock. 8B6T takes 8 binary digits and maps these to 6 ternary digits. Such a coding technique is capable of transmitting 2.6 times more data than a Manchester scheme using the same clock.

- **100Base-TX** uses two pairs of high quality Category 5 balanced UTP cable (or STP cable) and 4B5B coding with a high clock speed. This two pair configuration is the same as 10Mbit/s 10Base-T Ethernet, and the RJ45 plugs and 100m maximum segment length are also the same, but 100Base-TX requires higher quality cable. If 10Base-T Ethernet is installed with Category 5 UTP cabling, a later upgrade to 100Base-TX is possible.
- **100Base-T4** use works over lower quality Category 3 UTP cables, with a clock rate of 25 Mbits/s. While it achieves the same 100 metre segment length as 10Base-T and 100Base-TX, it uses four pairs of wires rather than two. CSMA/CD frames are multiplexed via the three data pairs while the final pair is used for control. 8B6T coding is used.
- **100BASE-FX** uses two strands of multi-mode optical fibre, one for transmission and one for reception. 4B5B coding is used, and the advantage of the 100Base-T implementations is that it allows larger networks.

5.6.5.3   Gigabit Ethernet

Gigabit Ethernet is a further extension to the IEEE 802.3 standard. In a similar process to Fast Ethernet, the basic operating principles of the frame structure and CSMA/CD principle have been retained while the operating speed has been increased, allowing Gigabit Ethernet to be used with the slower Ethernet standards. Figure 5.30 shows a typical example of Gigabit Ethernet use, where it is used as a backbone between 100Base-T hubs. The recently developed Gigabit Ethernet is replacing older technologies like FDDI in this application, as while the nodes must be replaced, it can use the same fibre optic links with significant speed increases. While there is no technical reason that Gigabit Ethernet could not be used for a terminal directly, there are few desktop computers requiring this sort of communication speed yet.

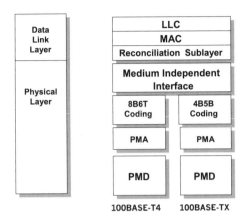

Figure 5.29   Fast Ethernet protocol stack

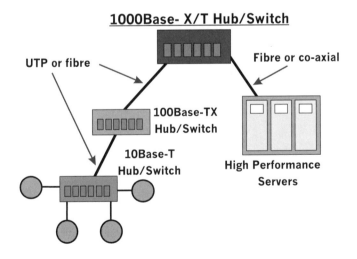

Figure 5.30   Gigabit Ethernet network

Four types of physical media are defined for Gigabit Ethernet.

- **1000Base-CX**: Media type is 150 ohm copper cord with a maximum device/hub distance of 25m.
- **1000Base-TX**: Uses 4 pairs of cat5 UTP – each transmitting at a rate of 250 Mbits/sec – with maximum separation between device and hub of 100m.
- **1000Base-SX**: Uses short wave-length transmission (850nm) over single or multi-mode fibre. Multi-mode fibre links can be up to 275m long, doubling to 550m for the single mode case.
- **1000Base-LX**: uses single or multi-mode fibre and 1300nm transmission. In this case, multi-mode fibre links can be 550m long, while single mode fibre links can be up to 5 km.

The access method associated with Gigabit Ethernet must be functionally equivalent to that associated with that of all other 802.3 versions. However, as noted in Section 5.3.1.3, CSMA/CD schemes require transmissions to have a minimum length equal to twice the propagation delay along the medium. Increasing the effective bit-rate tenfold compared to high speed Ethernet CSMA/CD and hundredfold compared to standard Ethernet CSMA/CD could be significant both in reducing transmission range or link efficiency.

The propagation/transmission delay constraint is addressed by ensuring that once a standard CSMA/CD frame has been transmitted by a terminal, that terminal continues to transmit 'extra' data to ensure that the terminal transmits data for at least the round trip delay time. Thus the terminal is able to detect any collision that may have occurred. This is known as carrier extension.

The extension data is added to frames that would be shorter than 512 bytes long (eight times the minimum for Ethernet at 10 and 100Mbit/s of 64 bytes). The nature of the data is not important and are only added after the FCS field. The properties of the physical coding use aid the switches/hubs in identifying and removing extension data. Note the CSMA/CD frame is unchanged and can be passed to the appropriate outgoing port without any change or extra processing.

Carrier extension has the disadvantage of adding a very significant overhead to short packets, to the extent that if fewer than the minimum number of bytes have to be transmitted, the frame is just as long as it was with 100 Mbit/s Ethernet. This is to be expected since the timing is arranged to be the same. To improve efficiency, if a terminal has more than one short frame to transmit, it can use the technique of frame bursting. If no collision has occurred within the minimum frame size of 512 bytes, the danger is past, and the terminal effectively controls the media. This means that a terminal can use carrier extension on the first frame, and if no collision has occurred, it is then permitted to continue to transmit its other short frames (see Figure 5.32). To prevent terminals from monopolising the media, the time any single terminal can transmit is limited. This limit is referred to as the burstlimit and is set to 8192 bytes.

5.6.5.4   10 Gigabit Ethernet

The pace of development of communication systems is illustrated by the current work to develop an extension to Gigabit Ethernet in the shape of 10 Gigabit Ethernet. This uses the same frame structure and protocol as Ethernet, although only switched access, referred to as full duplex, is possible. Full duplex was an option in slower Ethernet standards whereby

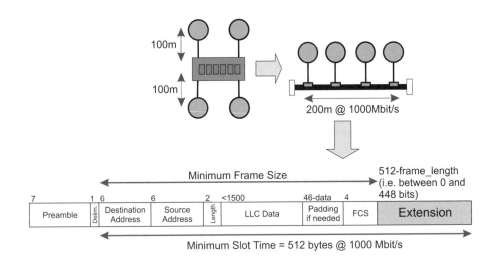

Figure 5.31   Collisions in a Gigabit Ethernet network

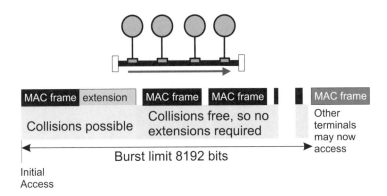

Figure 5.32   Frame bursting

the terminal has a dedicated connection to the switching hub, but will be mandatory for 10 Gigabit Ethernet. Its advantage is that it makes the full capacity of the medium available for a terminal and so utilisation can approach 100%, and since the medium is not shared, there are no collisions and so no restrictions on the frame size or need for carrier extension. At this speed most links would be point-to-point optical fibre anyway, with copper wires only possible for very short links under about 20 metres long. 10 Gigabit Ethernet is proposed as a lower cost alternative to SDH/SONET, the only other technology operating at such speeds.

5.6.6   IEEE Token Ring LANs

The IEEE 802.5 standard describes a token ring LAN. Unlike the approach of the IEEE 802.3 CSMA/CD LAN, in the token approach, access to the medium is controlled by use of a token. In token-based LANs, there is one token and nodes can only transmit while holding the token. Token networks operate in a logical ring (see Figure 5.33). Physically, ring, bus or star configurations are adopted. Although physical ring configurations are used in MANs, LANs usually have a bus or star configuration. A high speed version of the 802.5 standard requires a star configuration.

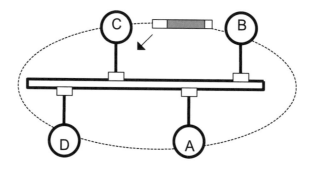

Figure 5.33   Token ring

In a similar way to the 802.3 standard growing out of the Ethernet LAN, 802.5 grew out of a token-based LAN developed by IBM. The system operates at 4 Mbit/sec and 16 Mbit/sec, and supports 8 priority levels. A 100Mbit/s version operating as a switched star is also defined.

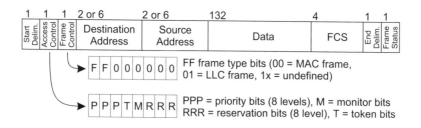

Figure 5.34   IEEE 802.5 frame formats

Token rings have an advantage over CSMA/CD LANs because access is deterministic and it is possible to allocate different priorities to nodes. However, a ring is not robust in one sense since the token passing is in a ring, and a node failure could break the ring and can cause the entire network to fail. Another LAN standard, the IEEE 802.4 Token Bus attempts to avoid this problem by allowing for nodes leaving or joining the logical ring. It supports 4 levels of priority, but is extremely complex with each node required to support 10 timers and 25 state variables. Tokens pass from high to low addresses with each node storing the address of the next station in the logical ring. If a successor node fails to accept a token (or is lost), the holder of the token commences a recovery procedure to find a new successor and remove that node from the logical ring. Nodes leave the ring by not accepting tokens. Periodically the current token holder will accept bids from nodes wishing to join the ring. This is achieved using special 'solicit successor' frames. The preceding node will then adjust its stored address of the next station.

5.6.7 IEEE 802.11 Wireless LAN

IEEE 802.11 defines a number of related wireless LAN standards which operate at different speeds. Like the other IEEE 802 series standards, the 802.11 WLAN standards share the other's link layer functions, making interworking of the different standards easier. As with the other standards in the series, the difference is in the transmission medium and configuration, and in the access method.

5.6.7.1 Medium Access Control

IEEE 802.11 WLANs use a contention-based scheme similar to CSMA/CD. However, as noted in Section 5.3.1.2, carrier sense schemes do not work well in a radio environment due to the hidden terminal problem. However, a slotted scheme would be difficult to arrange in a distributed WLAN environment since it would require a central base station, and in a local environment, the hidden terminal problem is not too severe, so IEEE 802.11 attempts a CSMA scheme.

Collision Detection in a radio environment is expensive, because it requires the terminal to be able to transmit and receive at the same time, and also that it be capable of distinguishing an incoming signal from the one it is transmitting (on the same frequency). This is a complex task, which if attempted, would make terminals expensive. In any case, if the collision was from a hidden terminal, it would not be detectable by the transmitter anyway. This means that the collision detection of the CSMA/CD scheme is replaced by 'collision avoidance' to give CSMA/CA.

Terminals listen to see if a transmission is in progress before they start to transmit. While this is not completely reliable, it does reduce the risk of collisions. If they hear no-one transmitting for a period of time, known as the Distributed Inter Frame Space (DIFS), then they transmit their packet. The receiving node will send an acknowledgement message after a short time (the Short Inter Frame Space, SIFS) after it receives the frame. These acknowledgements are required since the terminal itself cannot detect a collision, so a positive indication has to be sent back to the transmitter that the data has been correctly received. By waiting for the DIFS, other terminals will not mistakenly barge in during the SIFS and disrupt the acknowledgement of an otherwise correct message.

A terminal detecting a transmission on the channel when they want to transmit will wait as in the case of IEEE 802.3. However, unlike the 1-persistent case of IEEE 802.3 where a terminal would make its first attempt to transmit immediately the channel became clear, in

IEEE 802.11 protocol requires that terminals wait a random time even on this first attempt to reduce the probability of collisions.

The protocol also allows terminals to announce their intention to transmit. The transmitter sends a signal to the receiver requesting permission to send, and the receiver acknowledges this, after which the transmitter sends its data. By sending messages from both ends of the link, it is more likely that any terminal close enough to interfere with the transmission will also hear either the transmitter or the receiver's handshake, mitigating the hidden terminal problem.

5.6.7.2   Configurations

WLAN networks have an effective bus structure with the air interface forming the common medium. A common arrangement is to have a wired node attached to a LAN allowing access to the outside world or fixed peripherals like printers. Since the WLAN has no central controller, it is also possible to have an independent grouping, termed an *ad hoc* network, which is a network formed from any other terminals within range (see Figure 5.35).

Figure 5.35   Wireless LAN configurations

5.7   Connecting LANs

The original specifications for IEEE 802.3 CSMA/CD LAN set physical limits in terms of the length of a segment and the number of users that could be attached to that segment. For the most common version of CSMA/CD these limits are 185 m with a maximum number of users set at 30. Clearly an individual segment has limited value by itself – particularly in larger organisations. However, repeaters, hubs and bridges can be used to interconnect individual segments to effectively extend LAN coverage.

5.7.1   Repeaters

For the IEEE 802.3 CSMA/CD LAN, a maximum of 4 repeaters to be used to create a LAN equivalent to a 'single' segment 985 metres long capable of supporting up to 150 attached nodes. This creates a larger collision domain. The interconnection can either be done in cascade or in a multi-port fashion.

Since a repeater is a simple device – it repeats (broadcasts) incoming bits to all other enabled segments – the CSMA/CD protocol is completely oblivious to the presence of the repeater. This means that simultaneous transmission by any pair of nodes on any segment will result in a collision. The only extra functionality offered by the use of a repeater is the ability for a network manager to disable individual segments and stop the repetition of bits between that and the remaining segments. Such isolation can ease fault management, but must be done with care - for instance in the case of the cascade configuration, segment isolation can result in the 'large' LAN being split in two.

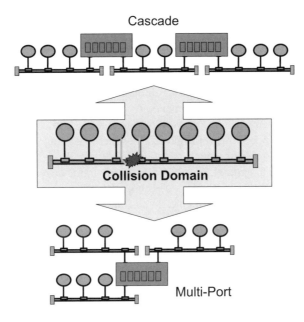

Figure 5.36   Connecting LANs with repeaters

Generally speaking, the role of a repeater is to provide physical interconnection of LAN segments. However, it should be recognised that, despite its simplicity, a physical bus-based structure is not ideal. Hubs are active devices that provide MAC functionality (typically CSMA/CD) in a star-based manner. The logical operation is still that of a broadcast bus but stations can now communicate via a central device using transmission media other than just co-axial cable, e.g. UTP5.

For 10BaseT RJ-45 connectors are used with a maximum distance of 100m allowed between device and hub – maximum node separation is limited to 200m. LAN segments can be further interconnected in a hierarchical fashion through the use of backbone hub (really a repeater).

Hubs and repeaters essentially perform the same function – emulating CSMA/CD. Hubs generally provide additional network management features including the means to isolate individual attachments and to gather and display segment statistics such as collision rates, utilisation, etc.

5.7.2   Bridges

Unlike hubs and repeaters, bridges allow individual LANs to retain their independence. Interconnecting three CSMA/CD LAN segments via a hub creates a single LAN with a maximum aggregate rate of 10 Mbit/s. Using a bridge allows the maximum aggregate rate to exceed this (30 Mbit/s) as there are essentially three separate collision domains. A collision in one segment will not be echoed in the other segments.

5.7.2.1 Transparent Bridges

The most common type of bridge is what is known as a transparent bridge or a spanning tree bridge. Such a device is intended to be completely open so that it can be used simply and quickly. Essentially it operates in a promiscuous mode. It accepts all frames transmitted on all the LANs it is attached to, and then attempts to forward the frame to the appropriate destination LAN by looking up a large routing table.

The routing algorithm works as follows. Initially the tables are empty and all frames received are 'flooded' on to all other segments. As frames are received by the bridge, it can work out which LAN each node is. For example, node A wishes to send a frame to node D via bridge 1 in the network shown in Figure 5.37. When the bridge receives the frame it is forwarded on to both LANs containing nodes B and D. The bridge also now 'remembers' which segment A is on and updates the table. Should any frame arrive at bridge 1 for node A, the bridge now knows what segment the node is on.

To ensure that tables reflect the dynamic state of the network (machines going on and off line) the tables are periodically purged and the process restarted. However, for this scheme to work, there must be no loops in the structure and a spanning tree can be formed.

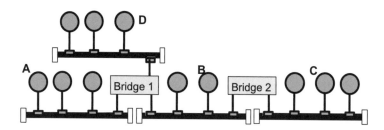

Figure 5.37   LAN segments connected by bridges

5.7.2.2 Spanning Tree Algorithm

The simple approach of not permitting any loops is not particularly robust. Should a bridge near the top of the structure fail, then much of the network will be unavailable. We can counter this risk but introducing loops – multiple connections between segments – but if we do that the simple algorithm described above will not work and packets could endlessly circulate. For this reason, an algorithm is built into the bridges that prunes the paths available into a spanning tree, a loop-free subset of the system.

A single root bridge is chosen for the LAN, the one with the highest priority and the lowest ID. Bridges regularly exchange messages allowing each bridge to determine which of its ports offers the least path cost to the root bridge. This port is called the root port and is used to communicate with the root bridge. Path cost associated with port depends on bit rate. The higher the bit rate, the lower the designated cost.

A designated bridge is chosen to forward frames from each of the LAN segments based on the least path cost to the root bridge from the LAN segment. If two bridges have the same path cost, then the bridge with the lowest ID is selected. The port connecting the designated bridge to the cable segment is called the designated port. After this is done, any bridge port which is neither a root port nor a designated port is switched off. This leaves only one root from any segment to the root bridge, forming a spanning tree. Bridges periodically exchange messages so that, should a bridge fail, the spanning tree will alter to take this into account.

This process is shown in Figure 5.38. The original network with multple connections is shown on the left. The network after the operation of the spanning tree algorithm is shown on the right, with root ports (RP) and designated ports (DP) marked.

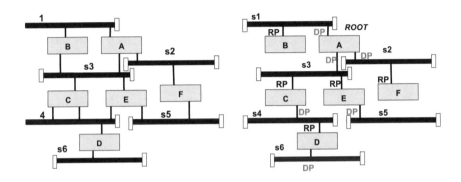

Figure 5.38   Spanning tree

5.7.2.3   Source Routing Bridge

The spanning tree approach is not optimal as only a sub-set (the spanning tree) of the network is used. An alternative to spanning tree-based transparent bridges are source routing bridges. Developed for use in token ring networks, this scheme ensures that the best possible use of LAN bandwidth is made.

Essentially the scheme relies on the transmitting node knowing where the destination node is and what path to take. This information is obtained by the use of broadcast 'discovery' frames which are transmitted whenever the node does not have such routing information. Routes are contained in the frame. This allows the system to find all the routes and choose the best, but the transmission of discovery frames can severely impact overall performance and the scheme is not transparent.

5.8   WAN Protocols

5.8.1   SDH/SONET

Prior to the introduction in 1992 of SDH/SONET, the fibre-optic systems were based largely on proprietary architectures and equipment. This meant that operators were often tied to a single equipment manufacturer for their switches, etc. Not unexpectedly operators were in favour of an open standard that would allow them to purchase compatible equipment from a variety of manufacturers, forcing more competitive equipment prices and reducing the need for expensive and complex inter-working with other networks.

A further disadvantage of the traditional systems is that they were hierarchical plesiochronous technologies. A plesiochronous system is one whereby different bit-streams exhibit variations in their bit-rates. SDH/SONET applies a synchronous approach whereby bit-streams have exactly the same bit-rates. The main disadvantage of a plesiochronous system stems from the fact that in order to multiplex several streams that have different bit-rates, additional bits are added into bit-streams in order to bring all streams up to the same common rate. This technique is known as bit-stuffing, but should not be confused with the similar bit-stuffing process used to delineate frames (see Sectionlink:frameing). In order to extract a particular bit-stream in a plesiochronous system, the entire signal hierarchy has to be demultiplexed, the remaining bit-streams have to be remultiplexed before being forwarded towards their destination; a complex and costly operation.

The SDH standard defines Synchronous Transport Modules (STMs) for carriage of the signals. STMs are defined for different bit rates, the lowest of which is an STM-1 frame. An STM comprises an overhead section and a payload section (administrative unit). The format of an STM-1 frame is 9 rows by 270 columns where each location holds a byte. Of these 270 columns, the first nine form the section overhead, with the remaining 261 holding the payload. The payload section holds Virtual Containers (VCs). VCs may be nested, so a VC may contain another VC. An STM-1 payload can hold one VC-4 which in turn can hold other VCs, e.g. a VC-4 can hold four VC-31s. Additionally VCs can be concatenated, providing a range of stream bandwidth options. The location of the VC within the administrative unit is not fixed and is permitted to 'float' over successive STMs; this allows compensation for phase misalignment. For example, if one of the tributary signals is running at a lower rate the VC can be permitted to slide progressively (change its start position within a frame) over a period of time. A key attribute of SDH is that pointers in the section overhead indicate the start of a VC (i.e. the start of a data package) within an STM. This allows data to be readily extracted from the bit-stream.

SDH has a comprehensive range of alarm and error messages (referred to as defects and anomalies). The alarm messages are transmitted in the SDH overhead section. Complete link failure is indicated by the Loss of Signal alarm. The transmitting node is informed of the link failure by the receiving node via the reverse link in the form of an Remote Defect Indication alarm. When the LOS alarm is generated, SDH's Automatic Protection Switching switches future transmission through a back-up link. Similarly, errors on the received signal are indicated by the Remote Error Indication alarm. The reliability features of SDH/SONET are a significant factor in its popularity.

5.8.2   ATM

IP is a very flexible protocol which works over a number of underlying networks and transmission technologies. However, that flexibility comes at a cost. The generic nature of the IP packet format means that they are not optimised to any transport mechanism and are more difficult to process. This is important for high speed networks where complexity translates to processing delays at switches and routers.

Asynchronous Transfer Mode (ATM) takes a different approach. It uses small fixed sized packets called cells, with only very simple functions being performed in the transit nodes. Some simple error detection is carried out on the header, but no error checking or recovery is performed on the cell contents, and connection oriented information transfer is used allowing very fast and simple hardware switching within the network.

Whatever the type of traffic, it is carried in a series of the cells. Low bandwidth traffic will use cells much less frequently than, say, high bandwidth traffic such as video. Since the use of cells is not synchronised, but on demand, the cell stream does not fit into a frame structure. When the traffic source is producing a heavy burst there will be many cells used, whereas when there is no traffic from the source the cells will be available for use by other traffic streams, i.e. statistical multiplexing.

5.8.2.1 ATM Services

ATM supports 4 traffic types, summarised in Figure 5.39:

- **CBR**: Constant Bit-Rate designed primarily to carry synchronous time-sensitive traffic such as voice.
- **VBR**: Variable Bit-Rate is divided into two sub-classes: real time (RT) or non-real time. VBR is for services where the data stream is not generated or transported at a constant rate, e.g. video. For RT services the delay constraints (jitter and absolute) are much tighter. VBR service is specified in terms of both mean and peak bit rates.
- **ABR**: Available Bit-Rate provides a best effort service and attempts to make use of the peaks and troughs associated with multiplexed VBR traffic. When there is more capacity, the ABR sources can transmit at a faster rate but, when the VBR load increases, then such ABR sources are made to reduce their rates. ABR is designed for applications where delay is not an issue. ABR BW can be bounded so that the user at least receives a minimum level of service.
- **UBR**: Unspecified Bit-Rate is a low cost low QoS service that provides no performance guarantees. Ideal for applications such background file transfer.

	CBR	**VBR**	**ABR**	**UBR**
Cell output	Fixed	mean & peak guaranteed	Network Controlled	Unspecified
Delay	Delay Sensitive	Delay In Sensitive		
Congestion Control	NO		Yes	No

Figure 5.39   ATM traffic types

There are 4 basic ATM Adaptation Layer (AAL) services (see Figure 5.40: AAL1–5,[2] each designed to support different applications over an ATM network. Each service is described in terms of its switching mode, bit-rates, delay tolerance and a sample application is given.

---

[2] Originally AAL 3 and AAL 4 were similar but distinct types, but it was decided to merge them at the AAL layer to form AAL Type 3/4.

AAL 1	AAL 2	AAL 5	AAL 3/4
Connection Oriented			Connectionless
Constant	Variable		
Delay Sensitive	Delay InSensitive		
Voice	Packet Video	IP X.25	SMDS

Figure 5.40 AAL services

5.8.2.2 ATM Protocol Stack

The ATM Protocol Stack is shown in Figure 5.41. ATM Layer is completely independent of the physical mechanism used for transmission. It sends data, passed down by ATM Adaptation Layer, to its intended destination. The Transmission Convergence sub-layer provides an interface to the transport medium. Two of its most important functions are cell delineation and Header Error Control generation.

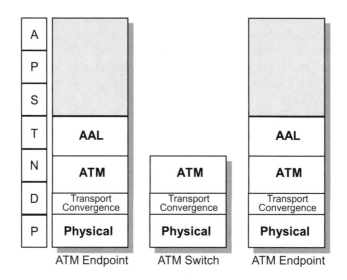

Figure 5.41 ATM protocol stack

Since ATM does not provide facilities like retransmission of corrupted data, ATM Adaptation Layer (AAL) gives an end-to-end protocol that provides the interface between the ATM layer and the higher layer protocols and applications. The AAL is responsible for accepting messages from higher level protocols and fragmenting them into smaller entities for transport into cells. It is also responsible for providing any additional services that might be expected by the higher layer applications such as timing synchronisation, sequencing, timing and error detection/correction. The AAL provides the functionality for ATM.

The fact the ATM has the AAL layer means that it is not really just a data link technology but can also provide the transport and network layers without the need for other protocols like TCP/IP. However, the popularity of IP in providing a common structure across networks has meant that end-to-end ATM has not proved as successful as was expected, although it has much better support for quality of services guarantees than the rather basic IP service. ATM is, however, commonly used to provide links within the core network.

5.8.2.3 ATM Transmission

ATM is a connection-oriented paradigm. Cells are transported over a transmission path and switched appropriately so that they reach their destination. The transmission paths are organised logically into Virtual Channels (VC) and Virtual Paths (VP). A VC is essentially a virtual circuit and is the basic transport unit of ATM – a connection between source and destination. Transmission order is guaranteed, subject to cell loss, over a virtual channel. A VP is effectively groups of VCs bundled together (see Figure 5.42).

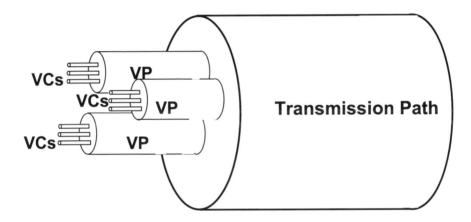

Figure 5.42 Virtual channels bundled in a virtual path

An ATM cell is 53 bytes long, with 48 bytes of information and a 5 byte header. The header is used for addressing, error control and network management. There is provision for a single bit to be used to indicate high or low priority. This feature has received considerable research attention in the development of priority schemes for call acceptance.

The two header formats are shown in Figure 5.43; the first represents the cell header for the User to Network Interface (UNI) while the second refers to the network to network (NNI) interface. The format and hence the functionality of these cells are similar. The only difference is that the NNI cell allows more addressing capabilities – 12 bit VPI – at the expense of the Generic Flow Control (GFC) Field that controls cell transfer when multiple access is needed to a medium.

5.8.2.4 ATM Traffic Contract

Prior to direct cell flow between end points, a connection must be set up. Part of this process is the negotiation of a traffic contract that describes the traffic characteristics to be adhered to by all parties. The key elements are a Source descriptor and QoS required. The descriptor will contain information about transmission rates, delays and loss.

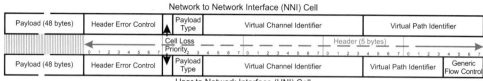

Figure 5.43   ATM cell formats

The Peak Cell Rate (PCR) represents the upper limit of the cell flow. This rate is not required all the time unless the traffic is CBR. The Sustainable Cell Rate (SCR) is upper limit to the mean cell rate and will be less than or equal to the PCR. Burst tolerance represents the fluctuations in cell inter-arrival times.

Cell Delay Variation (CDV) describes the differences that exist between the departure times (or arrival times at destination) of consecutive cells in a given flow.

Cell Loss Rate (CLR) defines the number of cells that are discarded within a circuit. Cell error rates refer to the corruption of cell contents which are mainly caused by poor failures in the transmission medium. CLR is more complete and will be greater than cell error rates.

The process of defining the traffic contract and then allocating resources to support a connection and meet QoS requirements (if possible) is known as Connection Admission Control (CAC).

5.8.2.5   Usage Parameter Control

Having agreed a contract between user and network, the network must then ensure that the traffic being placed into the network by a source does not exceed the limits of the contract. Breaking the contract can result in congestion which may affect other users who have not broken their contract. The contract may be broken maliciously by the user or network effects may inadvertently cause a flow to exceed agreed limits. There is therefore a need for policing traffic flows which is more generally known as Usage Parameter Control (UPC).

Using a policer, the network monitors cell flows (on a VC or VP) basis and identifies offending cells. These cells are either tagged (to be dropped at a later stage if they encounter congestion conditions) or dropped directly. Tagged cells may incur an additional cost to the network which may be passed on to the user. Tagging is achieved by changing CLP bit and part of AAL header. Conforming cells are all passed through unaltered.

The peak rate, and sometimes the mean rate, is achieved by using a 'leaky bucket' mechanism. This is a simple counter which increments as cells arrive and decrements at a pre-agreed rate, in a similar manner to a bucket with a hole in it. If counter reaches a pre-defined threshold, cells are tagged or dropped. Policers are located both within the network and at the edge.

Sources may shape traffic prior to entering the network to ensure that contract is not broken and to maximise performance and similar devices are used to achieve this effect.

5.8.2.6   ATM Switching

ATM is based on the concept of high-speed hardware-based switching devices (see Figure 5.45). There are a variety of different switching architectures available, each with

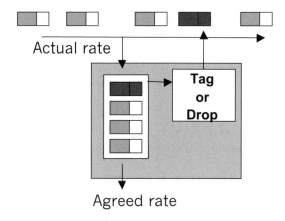

Figure 5.44   ATM policing

their own pros and cons. Regardless of their internal structure, an ATM switch has two key functions:

- It must be capable of recognising and interpreting the addressing information within a cell's header. Along with the incoming port, it is this information that indicates where the cell is to be routed.
- The ATM switch must transfer the ATM cell from its input to the correct output very quickly. There is the additional requirement that cell headers must also be updated prior to leaving the switch.

Routing is achieved either by a form of self-routing or table-based routing. In the former, additional information is added by the input stage of the switch to the cell header which is used to direct the cell through the switch fabric. The header is removed prior to cell departure. In table-based routing, look-up tables are used to map address header information to paths through the switch fabric and associated elements.

   ATM switches, such as that shown in Figure 5.45, are essentially multiplexors. They must deal with contention, the situation when two cells arriving at the same time require to leave via the same port. This is resolved by buffering or discarding. Since both affect QoS – the former by adding delay, the latter by increasing the cell loss rate – they must be controlled.

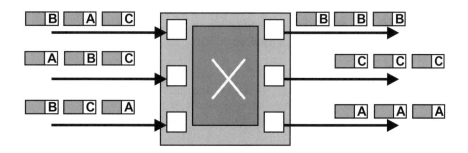

Figure 5.45   ATM switching

The ATM cell contains two address fields, the Virtual Channel Indicator (VCI) and the Virtual Path Indicator (VPI). The VCI is used on a link-by-link basis to determine the channel over which the cell is to be passed. A routing table in each node will convert the VCI on the incoming cell to the appropriate VCI for the cell to be sent onto its next node. The VPI is used to tell the system over what end-to-end connection the cell is intended for. Virtual path switching is also possible, and in this case both VPI and VCI fields are used. Switches are required to do address translation and modification of VPI and VCI fields.

VP and VC switching is shown in Figure 5.46. In VP switching, the paths, and all the VCs they contain, are switched. In the VC switching, the individual VCs are switched,

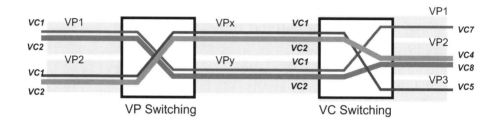

Figure 5.46   Virtual path switch

The term virtual is used because while the cell sees the channel or path as its own, it is not dedicated to that cell alone, but it is in fact shared by many other cells. Allocation of resources to each VP or VC is a critical management issue.

Different connection types are possible. Permanent Virtual Circuits (or Connections) (PVC) are equivalent to Cross Connects, and allow a long-standing allocation of resources associated with a path or flow. Switched Virtual Circuits (SVC) switch paths and hence resources dynamically as and when a connection is required.

It is not practical to assign VC numbers on a network-wide basis. Instead VC numbers are allocated on a 'hop-by-hop' basis. Hence a particular connection between source and destination will be known by different VC numbers as it traverses switching nodes through the network. VC identifiers are dynamic and it is the job of network layer to manage the change. The network in Figure 5.47 shows three routes plus the routing table for node X.

5.8.2.7   ATM Cell Fields

The ATM cell format is shown in Figure 5.43. In addtion to the 48 byte payload, there are three main control fields. The Payload Type Identifier (PTI) indicates whether a cell refers to either a user or network flow, whether the flow is congested or not, plus what kind of AAL the cell relates to.

The second control field is the Cell Loss Priority (CLP) which is used by UPC to determine the relative priority of a cell. If the CLP is set (= 1) it indicates that cell has a lower priority and can be discarded by the network if congestion occurs. A CLP of 0 indicates a higher priority flows. The CLP can be set by source or modified by the network.

The 8 bit Header Error Control field protects the cell header (to a degree) from any errors that may result in incorrect routing. The payload is not protected here - this is left to the upper

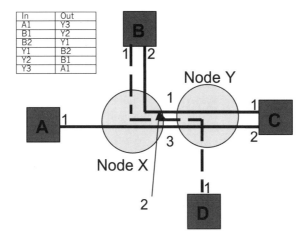

Figure 5.47   Virtual path numbering

layers or to the application directly. Worthwhile payload protection would require more than 8 bits and would increase cell overhead.

Header protection is achieved using an 8 bit generator polynomial that is capable of detecting and correcting all single bit errors plus detecting (but *not* correcting) all 2 bit errors within the 4 byte functional part of the header. The detection and correction mechanism is based upon two polynomials:

$$G(x) = x^8 + x^2 + x + 1(100000111)$$
$$C(x) = x^6 + x^4 + x^2 + 1(01010101)$$

This can be achieved with simple digital circuits on a bit-wise basis.

Protection operates in two modes, correction and detection, as shown in Figure 5.48. Normal operation is in correction mode where unerrored cells are passed on to higher levels. If a single error occurs, then the error is corrected and the cell is passed on but the state changes to the detection state. State change also occurs when multiple errors are detected. The device moves from the detection state back to correction state when an unerrored cell is detected.

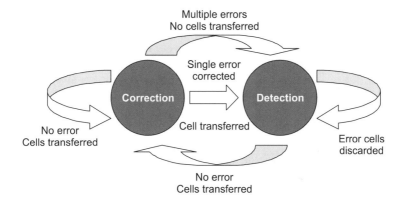

Figure 5.48   ATM error detection

The synchronisation procedure is shown in Figure 5.49. The system starts in the search state and only moves into the pre-synch state when a successful HEC process is obtained. It will then either return to search if the next HEC process fails, or move to the Synchronised state (where we want to be) when Δ successive HECs occur. The system will remain in that state until α successive HEC fail, whereupon it will return to the search state. Typically α and Δ are 7 and 6, or 7 and 8. There exact value depends upon the transmission scheme used.

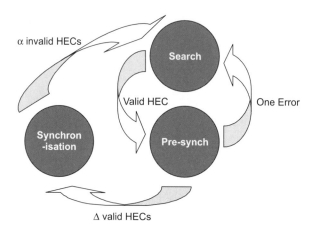

Figure 5.49   ATM synchronisation

5.9   Transporting IP over WANs

5.9.1   Point-to-Point Protocol

Point-to-point protocol (PPP) is a protocol for the transport of network packets over a point to point link. It was designed to replace SLIP (Serial Line IP.) Unlike SLIP, PPP can carry a range of network layer datagrams whereas SLIP is designed exclusively for serial transmission of IP. PPP was initially devised for communications over simple serial links. Due to its configurability and reliability, it is very likely to be the protocol used over a telephone line to connect a computer with dial-up access. However, PPP is not restricted to any particular transport technology and can operate over any point to point link. It is used, for example, for transporting IP (and other layer 3 datagrams) between routers over SDH/SONET/.

The frame structure is shown in Figure 5.50. Every PPP frame begins and ends with the string 01111110. This is used to delimit the frame. Since the frame has a variable length, it is essential that this string does not appear as part of the information block. PPP uses byte stuffing, which is similar to the bit stuffing described in Section 5.2 but which operates at a byte level. If 01111110 does appear in the message, it is preceded by the escape byte 01111101. The data byte which is escaped is XORed with 00100000. If the escape sequence itself appears in the message, it too is preceded by the escape byte. The escape byte is always removed from the message data by the receiver. Its only purpose is to prevent false interpretation of message data as a frame delimiter.

The address and control fields both have 8 bits and are always set to 11111111 and 00000011 respectively, although the standard does not rule out the possibility of other values being defined at a later date. The protocol field defines the network layer protocol carried by the frame.

1	1	1	1 or 2	*variable*	2 or 4	1
Flag 0111110	Address 11111111	Control 00000011	**Protocol**	payload	Checksum	Flag 0111110

Figure 5.50 PPP frame

PPP uses a Link Control Protocol to set up a connection and negotiate parameters for transmission. LCP is also used to check the status of a connection and bring it down in an orderly manner when the session is over.

The framing aspects of PPP come from the HDLC (High Level Data Link Control) protocol which PPP uses as a basis. HDLC has the same frame structure and delimiters, which is where the rather redundant address and control fields come from. HDLC, which was developed in 1979, is a more complex protocol than PPP. When HDLC was developed, the problem of timing and synchronization between devices was greater than it is at present due to more advanced processing techniques. HDLC is asynchronous in that the frame delimiter can occur at any point and be detected. The problem with this approach is the need for byte stuffing and the problem with errors corrupting a packet and causing a loss of synchronisation.

PPP can work with other data link protocols and another possibility is SDL (Simple Data Link) protocol. SDL has a frame length field in the header which identifies the length of the frame and allows faster processing of the packet. The frame length field is protected by a frame check sequence against errors.

5.9.2 IP over SDH/SONET

SDH/SONET is a simple link layer interface and thus is not able to transport IP on its own. However, PPP/HDLC can be used as a data link protocol to allow IP transport. Proposals are also being worked on to allow PPP/SDL to be used.

ATM can also use SDH/SONET for its transport. In this case ATM would carry IP as described in the following section, and SDH/Sonet would carry the ATM cells. The protocol stacks for the different approaches are given in Figure 5.51.

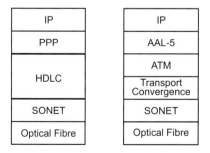

Figure 5.51 Alternative approaches for carrying IP over SDH/SONET

5.9.3   IP over ATM

AAL5 is used to carry IP PDUs over ATM networks. AAL5 is simple and involves minimal overhead in the transport of higher level PDUs as ATM cells.

Essentially, the IP datagram is passed down as the payload of the AAL 5 Common Part Convergence Sub-layer (CPCS) PDU. The CPCS payload varies between 0-65538 bytes, so any size of IP packet can be carried, and an extra PADding field is added to ensure that the PDU is an integer number of 48 bytes so that it fits within the ATM cell payload – note that AAL5 introduces no AAL overhead and all 48 bytes are available to carry information. The CPCS PDU also contains a 2 byte length field and a 4 byte CRC field.

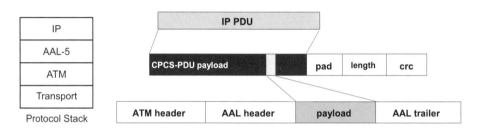

Figure 5.52   Carrying IP packets over ATM

5.9.3.1   ARP for IP over ATM

ATM/IP interworking enables IP routers to use cell-based channels to provide high speed interconnectivity. Connections between routers can be in the form of SVCs or PVCs. In addition to PDU formatting, ATM/IP requires the use of address resolution, to match IP addresses of routers with appropriate ATM addresses. The procedure is similar to LAN-based ARP except recognise that ATM is connection based rather than broadcast based. Each router would maintain an ARP table – essentially a dynamic list – which would be consulted each time an IP packet required to be forwarded. If the router table does not contain the appropriate address, the ARP procedure is invoked. There are essentially two forms: ARP server and Broadcast ARP. In the former, the router forwards an ARP request over a VC towards the server who then responds with an appropriate address for the router, which is then able to forward the IP datagram. In the broadcast ARP system, the router forwards an ARP request to all routers associated with the ATM network over a dedicated VC reserved for ARP requests. The destination router matches its IP address with the ATM address and replies to the source router using the same VC. The source router can then transmit the IP datagram.

5.10   Metropolitan Area Networks

Metropolitan Area Networks are high speed networks over a larger area than would be served by a LAN, but not over the long distances found with WANs. Unlike LANs, MANs are run by telecommunication companies and serve a number of different organisations.

The ideas behind MAN development were derived from LANs, and MANs are often described as large and fast LANs. However, this does not mean that LAN MAC schemes are directly suitable for use within the MAN environment.

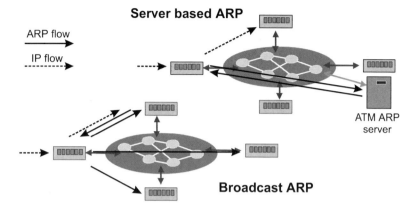

Figure 5.53   ARP for ATM

One requirement associated with MAN systems is bandwidth, i.e. higher bit-rates. To support the range of applications that are typically associated with MAN technology requires high speed data links; 100 Mbits/sec and above. MAN systems are also required to cover distances greater than that associated with LAN systems; perhaps 100 km rather than less than 5 km. Finally, it is necessary to consider the nature of the services supported by MAN systems. Unlike LAN systems, MANs will be required to carry synchronous as well as asynchronous traffic; video and voice services, for example, represent constant bit-rate source which is in contrast to more traditional asynchronous data traffic.

The IEEE 802.3 CSMA/CD LAN standard is unsuitable for use in a MAN environment for a number of reasons:

1. Increasing the bit-rate of the link would mean that the protocol would become even more inefficient. The minimum frame length would need to be increased to ensure no increased probability of collisions.
2. Increasing the length of the bus would increase propagation delay and thus increase the probability of collisions. This, of course, could be compensated for by increasing the minimum frame length but at the cost of further transmission inefficiencies.
3. CSMA/CD is essentially an asynchronous access mechanism and is more suited to carrying asynchronous data traffic. It is far from ideal when supporting synchronous traffic which requires constant and predictable delays.

Reserved access is more appropriate for these larger networks. This allows more efficient use to be made of the medium, which is important as the cost of a large network depends more on the length of the links than on the cost of the individual nodes themselves. Trading off some node complexity for improved efficiency is a good deal. Another important factor is that since the network is used by a number of organisations and run by a third party, access to the medium must be guaranteed. Contention-based schemes do not allow this.

The dividing line between a MAN and a LAN has always been somewhat fuzzy, and FDDI, for example, is often used in a LAN environment. As LANs get faster, developments like Gigabit Ethernet offer MAN-like capabilities at lower cost. At the other end of the scale, WANs are now able to offer data rates up to and even exceeding MAN rates with no compromise on distance. It is therefore likely that any new technological developments would be targeted either at the WAN or LAN market.

5.10.1   Fibre Distributed Data Interface

Fibre Distributed Data Interface (FDDI) was developed by ANSI (American National Standards Institute) purely as a high-speed data network. A second generation version, FDDI II, evolved to provide integrated traffic capabilities and it is this version that will be discussed.

FDDI is a high-speed (100Mbit/s) network which utilises a physical ring topology. In a MAN setting, its key features include:

- A token-based MAC scheme, based upon the IEEE 802.5 standard, which is compatible with the IEEE 802.2 LLC standard. FDDI is compatible with all IEEE LANs.
- A dual ring topology which allows a total fibre path of 200 km (100 repeaters with a maximum of 2 km between them) and a maximum of 1000 fibre connections. As a result, FDDI can support up to 500 users linked over 100 km. Fibre can be multi-mode or single mode. FDDI can also be used as a LAN standard with twisted pair wiring replacing the fibre. The maximum distance between repeaters then falls to 100m. A common use for FDDI is to link together LANs as shown in Figure 5.54.
- The ability to dynamically allocate bandwidth so that both asynchronous and synchronous services can be provided simultaneously. FDDI supports up to sixteen 6.144 Mbit/s channels of circuit switched isochronous data streams. Each of these 6.144 Mbit/s streams is equal either to 3 CEPT (2.048 Mbit/s) frames or 4 T1 (1.544 Mbit/s) frames.

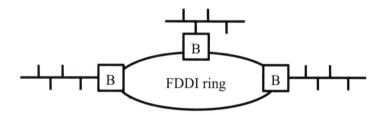

Figure 5.54   Using FDDI to link LANs

FDDI is based upon a dual ring topology. The two rings are used either for concurrent transmission, giving an effective bit-rate of 200 Mbit/s or, more normally, with the second ring as a stand by. In this case, the effective data transmission rate is 100 Mbit/s. This dual ring topology provides a high degree of fault tolerance. Should a node or link fail, then the two rings fold to form one ring (see Figure 5.55. It one break occurs, then full connectivity is maintained but the ring is twice its original length. Should a second break occur, then the network fragments into two separate networks, each able to operate independently of the other but with the loss of full connectivity.

Despite its name, FDDI can be used over a copper wire as well as over an optical fibre, although in this case it is really a LAN technology with a range between repeaters of only 100m. With an optical fibre, this distance rises to 2km, FDDI uses an actual bit rate of 125 Mbit/s per ring and a '4 out of 5' 4B5B encoding scheme whereby a symbol is represented by 5 clock periods and effectively represents 4 bits. This requires less bandwidth than Manchester encoding but, unlike Manchester encoding, no clocking information is present in the signal so a long preamble is used. The copper based version uses MLT-3 coding.

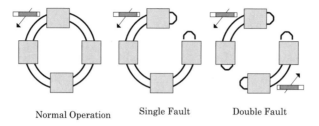

Normal Operation   Single Fault   Double Fault

Figure 5.55   Fault tolerance features of FDDI

Frame format allows variable length information field. Frames start with an eight byte preamble with is used to synchronise the receivers. This is followed by a single byte Start Delimiter (SD), which indicates the start of the frame, and a Frame Control (FC), which indicates the function of the frame.

FDDI supports either 16 bit or 48 bit addresses, and a mixture of address formats can be used on a ring. The Destination Address (DA) and Source Address (SA) can therefore be either 2 or 6 bytes long. The information field is followed by a Frame Check Sequence (FCS), and a 4 bit (i.e. one symbol) End Delimiter (ED), indicating the end of frame. The frame is followed by a Frame Status field, used to reply to the source. This can indicate that an error was detected, that the destination recognised its address, and that a frame was received.

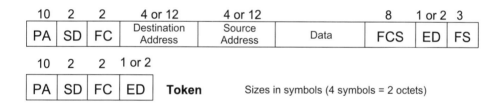

Figure 5.56   FDDI frames

5.10.1.1   FDDI MAC Operation

FDDI uses a token system similar to IEEE 802.5, and like that scheme allows for traffic to have different priorities. A station can only transmit frames if it has captured a token. There can be more than one token on the ring at once. The rules associated with token capture and release are determined by traffic type, token type and priority level. Token reception and capture are not synonymous and if a station receives a token but it is not allowed to transmit, it must return the token to the next station on the ring. The rules associated with token capture and holding are described by the Target Token Rotation Protocol. The FDDI system supports 4 different traffic types – each with a different priority level. These traffic types are split between two synchronous data types and two asynchronous traffic types.

The two types of synchronous traffic are termed synchronous traffic and isochronous traffic. Isochronous traffic is circuit-switched synchronous data which is synchronised to a single (normally external) time reference. The traffic is carried using special frames whose length is fixed for the duration of the connection and is determined by the capacity required. The timing for this traffic is based upon 8kHz cycles and is derived and controlled by a station known as the Cycle Master. Stations negotiate for the use of this capacity and the right to be the Cycle Master.

Bandwidth for both forms of synchronous traffic is allocated from the total FDDI bandwidth. The remaining bandwidth is used to carry asynchronous traffic of which there are two types:

- **Restricted Asynchronous traffic**: This traffic can be transmitted upon the capture of either a restricted or unrestricted token. Stations negotiate (using unrestricted tokens) to transfer information in a restricted form. This is essentially a virtual circuit or connection-oriented mode, where stations compete for the remaining bandwidth on a per-call basis.
- **Non-Restricted Asynchronous traffic**: This traffic can only be transmitted upon the capture of a non-restricted token. Essentially a datagram or connectionless mode, where stations compete for the remaining bandwidth on a per packet basis.

In addition to the two classes of asynchronous traffic there are 8 levels of priority which can be used if desired.

The token scheme for FDDI differs from that of IEEE 802.5 in that a token is released immediately after data is transmitted rather than when an acknowledgement is received. This has the advantage that the utilisation of the medium is increased, but presents the problem that there may be more than one token in the ring at one time. This means that priority scheme of IEEE 802.5 will not work because a station could reserve capacity with one token only to see it used by another station on the basis of a different token. Another issue with FDDI is the presence of synchronous data.

The FDDI scheme aims to allow all nodes to transmit their synchronous data while sharing the remaining capacity fairly among nodes for their asynchronous traffic.

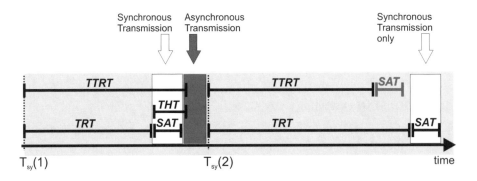

Figure 5.57   TTR protocol

During an initialisation phase, a reference time called the Target Token Rotation Time (TTRT) is agreed by all the nodes. This is the time, under load conditions, that it should take for the token to circulate around the ring. The TTRT is such that it is greater than the synchronous data requirements of all the nodes (their synchronous allocation time, or SAT), plus the frame time, the token time, and the propagation time round the ring (the ring latency (RL)). If this were not the case, then not all nodes would be able to transmit their synchronous data. The difference between the TTRT and the sum of all the synchronous data requirements

($\sum$SAT) is the time available for asynchronous transmission. The difficulty is for nodes to work out how long they may transmit given that the asynchronous data requirements of all the nodes are dynamic.

In order to work out how long it has to transmit, each node maintains a timer, called the Token Rotation Timer (TRT). This counter counts how long the token has taken to go round the ring and come back to that node. If, when the token is received, the TRT counter has not yet reached the TTRT, then there is 'spare' time to transmit asynchronous data. The amount which can be transmitted is the difference, i.e., TTRT − TRT. This difference is termed the Token Hold Time (THT). On receiving a token, a node will transmit its synchronous data (even is more than the TTRT has elapsed), and then, if TRT < TTRT, asynchronous data up to the maximum of THT or as much asynchronous data as is available, will be transmitted.

5.10.1.2   FDDI Performance

The FDDI protocol has two important characteristics:

- A guaranteed maximum response time. Under worst-case conditions, the arrival of two successive frames from a synchronous source will not exceed 2 × TTRT. This would occur if a node received the token back immediately (so TRT = 0), whereupon it would transmit for THT = TTRT − TRT = TTRT. The remaining synchronous frames will take at most TTRT to transmit (since TTRT > $\sum_i$SAT$_i$)
- The maximum network utilisation is predictable. It is:

$$\text{Utilisation} = \frac{N(\text{TTRT} - \text{RL})}{(N \times \text{TTRT}) + \text{RL}} \rightarrow \frac{\text{TTRT} - \text{RL}}{\text{TTRT}}$$

where RL is the Ring Latency and N is the number of active stations. It can therefore be seen that FDDI is sensitive to ring size, the value of TTRT and the number of stations (total and active). Increasing TTRT leads to increased efficiency, but increased delay to synchronous frames. Larger rings have increased Ring Latency, thus reducing efficiency.

5.10.2   Distributed Queue Dual Bus

IEEE 802.6 is a Distributed Queue Dual Bus (DQDB) standard for Metropolitan Area Networks. Although it is designed to serve similar needs to that of FDDI, it is very different in nature to the earlier MAN standard. DQDB was developed specifically to carry multiple traffic types which is in contrast to FDDI which originally supported only data traffic and evolved into FDDI II supporting asynchronous and synchronous traffic. Key features of DQDB include:

- Use of a dual bus architecture with each bus operating independently of the other. A looped bus topology can be used to provide a degree of fault tolerance.
- Data rates are variable (from 34 Mbits/sec to 155 Mbits/sec and above) and there is no single specified media. Coaxial cable, optical fibre and microwave paths can all be used.
- The operation is basically independent of the number of stations.
- Support of isochronous and asynchronous traffic coupled.

Although the system is called a bus, it is not a physical bus in the true sense. The bus is formed from a number of unidirectional point-to-point links between nodes (see Figure 5.58).

By having two buses, one in each direction, data can be sent to nodes on either direction of the transmitter. At the start of each of the buses is a head of bus station. This can be a separate entity or could be incorporated in the first node.

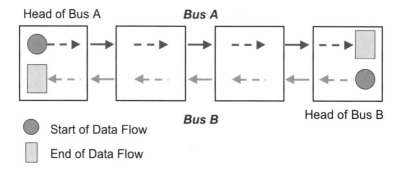

Figure 5.58   DQDB network configuration

The bus concept comes about because the data propagates to all nodes (at least all nodes 'downstream' of the transmitter). Nodes cannot remove data from the bus – they can only add it by changing a passing '0' into a '1'. The configuration of each node attachment is shown in Figure 5.59. The unit is such that nodes can monitor the individual bits within slots and modify them on a bit-by-bit basis before passing the data to the next node on the bus. The OR function means that a node can only modify the data by changing a '0' to a '1' but not a '1' to a '0'. Having simple addition circuitry like this means that the bus can operate very quickly, and while at first sight it may seem inefficient to pass data beyond the node for which it is intended, in practice to make use of this additional capacity would require complex resource management techniques and signalling which would not make any gain worthwhile.

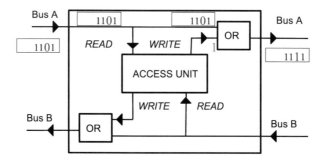

Figure 5.59   DQDB node operation

As with other IEEE 802 standards, DQDB retains full compatibility with the IEEE 802.2 LLC standard. The IEEE 802.6 standard defines a high speed network using the DQDB MAC and describes two protocol layers; the Physical Layer and the DQDB Layer. A simplified version of the DQDB protocol architecture is shown in Figure 5.60. The standard currently supports a range of transmission systems but does not explicitly specify a maximum bus length or maximum number of stations. These characteristics will be dependent upon the transmission system actually used.

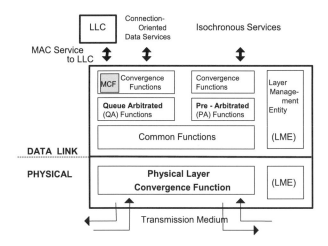

Figure 5.60   DQDB protocol stack

The DQDB layer is equivalent to the lower portion of the Data-Link Layer, i.e. IEEE 802 MAC layers. The DQDB layer provides support for three basic higher layer services: Synchronous traffic, Connection-oriented data traffic and Connectionless data traffic. The higher layer services are supported through the use of two different access methods:

- **Pre-Arbitrated Access:** This is used for time-sensitive services, i.e. synchronous traffic. This is achieved by assigning, for a particular connection, specific octet positions within particular slots in advance.
- **Queue Arbitrated Access:** This is used by the asynchronous sources who request the use of free un-assigned slots. Access to these slots is determined by the use of the Distributed Queue Access protocol.

DQDB has a synchronous frame structure based upon fixed length slots shown in Figure 5.61. Each slot has an Access Control Field (ACF) followed by a segment which carries the data. The Access control field contains the following elements:

- **Busy**: indicates whether the slot contains data or not.
- **Type**: indicates the type of the slot – synchronous or asynchronous data.
- **PSR**: Previous segment Received – essentially an acknowledgement.
- **Reservation**: Confirms that a slot has been reserved.
- **Request**: used by nodes to attempt to request access to a slot.

When a node has asynchronous data to transmit, it will use the bus going in the opposite direction to reserve a slot to carry this data. In this context, that bus is called the request bus while the other is called the transmission bus.

The node initialises two counters – request and countdown. The node then monitors the frames carried by the Request bus. The node informs the Head of Bus that it wishes to obtain a slot, by attempting to modify the request bit in the slots passing on the Request bus. However,

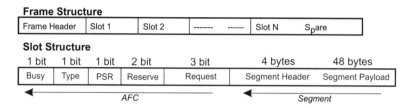

Figure 5.61 DQDB slot and frame structure

nodes downstream may also be requesting slots and request bits in passing slots may already by set. For every previously reserved slot that passes, the request counter is incremented. Thus, the node attempts to estimate how many nodes are ahead of it in the queue for a free slot.

To transmit on bus B:

inc **request** if Request bit set on bus A
dec **request** if Busy bit clear on bus B

When the node successfully alters the appropriate request bit (i.e. an unrequested slot passes), the contents of the request counter are transferred to the countdown counter. The node now monitors the slots passing on the Transmit bus.

request → countdown, set request to 0
dec countdown if Busy bit clear on bus B
when countdown = 0, use next slot with Busy bit clear on bus B

The queued access protocol will provide fair access to all nodes connected to the network if the maximum distance between nodes is less than the time it take to transmit a single 53 bit slot. However, if the distance between active nodes exceeds this distance, the QA protocol performance is reduced and there are problems associated with fairness. In the QA protocol, due to propagation effects, nodes closer to the head of bus will be able to seize free slots ahead of nodes further down the bus. This leads to unfairness whereby nodes close to the head of bus have lower waiting time than nodes in the middle of the bus. To counteract this, a bandwidth balancing mechanism is introduced which essentially divides up the available bandwidth and prevents individual asynchronous sources from exceeding their allocation. Although ensuring more fairness to asynchronous users, bandwidth balancing can result in wasted capacity.

5.11 Questions on the Link Perspective

5.11.1 Questions on Linear Codes

1. What is the probability that a 4 bit message will be received correctly in a system using a (7, 4) Hamming code if the probability of a bit error is 2%?
2. Which is more likely: more than one error when transmitting 16 bits through a BSC (Binary Symmetric Channel) with transition probability 0.1, or more than two errors when transmitting 8 bits through a channel with transition probability 0.2?
3. What is the minimum possible number of parity bits theoretically required for a two error correcting code which is 16 bits long?

4. Put the following matrix into standard echelon form
$$\begin{pmatrix} 0 & 0 & 0 & 1 & 0 & 1 & 1 \\ 0 & 0 & 1 & 0 & 1 & 1 & 0 \\ 0 & 1 & 0 & 1 & 1 & 0 & 0 \\ 1 & 0 & 1 & 1 & 0 & 0 & 0 \end{pmatrix}$$

5. Form the generator matrix of a $(15, 11)$ Hamming code and its parity check matrix.

6. The codeword 0110010 was received in a system using a Hamming code with the following generator matrix. What was the transmitted message?

$$\begin{pmatrix} 1 & 0 & 0 & 0 & 1 & 1 & 0 \\ 0 & 1 & 0 & 0 & 1 & 0 & 1 \\ 0 & 0 & 1 & 0 & 0 & 1 & 1 \\ 0 & 0 & 0 & 1 & 1 & 1 & 1 \end{pmatrix}$$

7. A binary coding scheme for messages is constructed from two information symbols b_1 and b_2 and three parity symbols (p_1, p_2 and p_3). The generator matrix is as follows:

$$G = \begin{pmatrix} 1 & 0 & 1 & 1 & 1 \\ 0 & 1 & 1 & 0 & 1 \end{pmatrix}$$

   (a) How many codewords are there? What are the codewords?
   (b) What is the minimum distance of the code?
   (c) What is the theoretical error correcting capability of this code?
   (d) Determine the parity check matrix H.
   (e) What is the syndrome for the following error patterns: (01000), (00101), (10010), and (11111)?
   (f) A code word (11010) is generated, which is distorted with the following error pattern: (10010). If the code is used for error correction, what will the decoded codeword be? Explain your choice.

8. A binary code is constructed by adding three parity check symbols to three information symbols (denoted by b_1, b_2 and b_3). For these code symbols:

$$\begin{aligned} c_1 &= b_1 \\ c_2 &= b_2 \\ c_3 &= b_3 \\ c_4 &= b_2 + b_3 \\ c_5 &= b_1 + b_3 \\ c_6 &= b_1 + b_2 \end{aligned}$$

   (a) What are the codewords?
   (b) Determine the generator matrix G and the parity check matrix H.
   (c) What is the minimum Hamming distance of the code? What is its error correction and detection capability?

9. A binary linear block code is constructed from a single parity check (SPC) code where for every three information symbols one parity check symbol is added so that there is an even number of 1s in each code word.

   (a) Determine the G and H matrices for this code.
   (b) What is its minimum distance?

10. A binary linear code has the following parity check matrix.

$$H = \begin{pmatrix} 1 & 1 & 1 & 0 & 1 & 0 & 0 \\ 0 & 0 & 1 & 1 & 1 & 1 & 0 \\ 1 & 0 & 0 & 0 & 1 & 1 & 1 \end{pmatrix}$$

(a) What are n and k for this code? How many codewords does the code have?
(b) By expressing the H matrix as $[P^{\top} I]$, where I is the identity matrix, give a systematic generator matrix (i.e. in the form $[IP]$, so the first k bits of the codeword are the same as the message).
(c) List all the codewords.
(d) What is the minimum distance of the code? How many errors can it correct?
(e) After transmitting two codewords through a binary symmetric channel using the systematic form of the code, (1000001) and (1001100) are received. Calculate the syndrome for each of these received codewords. Determine the most probable transmitted codeword in both cases, and from this the most likely message.
(f) In this example, assume that the binary symmetric channel has a transition probability of < 0.5. What does this say about the relative probabilities of errored and error free bits? If the transition probability was known to be more than 0.5, is there a simple way of using codes like this one, and if so, how?

5.11.2 Questions on Convolutional Codes

1. A convolutional encoder has the following encoding circuit.

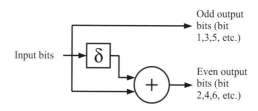

(a) Draw the code trellis for this code.
(b) What is the constraint length of the code? For this code, the minimum non-zero path is generated from the input 100... Find the minimum distance for the code, and state its error correcting capability.
(c) By repeating the trellis diagram for each input bit, and clearly showing each change in state, encode the sequence 1 0 1 1 0 1 0 0.
(d) By drawing the trellis diagram showing the path weights at each step, use the Viterbi algorithm to perform maximum likelihood decoding on 00 11 11 11 10 10 00 00. Hence state the most likely transmitted sequence.

2. A convolutional encoder has the following encoding circuit.

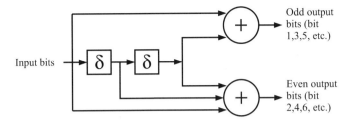

(a) Draw the code trellis for this code.
(b) What is the constraint length of the code? For this code, the minimum non-zero path is generated from the input 100... Find the minimum distance for the code, and state its error correcting capability.
(c) By repeating the trellis diagram for each input bit, and clearly showing each change in state, encode the sequence 1 0 1 1 0 1 0 0.
(d) By drawing the trellis diagram showing the path weights at each step, use the Viterbi algorithm to perform maximum likelihood decoding on 00 01 10 01 11. Hence state the most likely transmitted sequence.

5.11.3   Questions on LANs and MANs

1. A LAN based upon IEEE 802.3 CSMA/CD 10Base2 is used to carry data between two computers located/cabled 15m apart.

(a) If a simple connectionless service is employed by LLC, how long will it take to transfer a 1 Mbyte file between these devices if no other devices are using the network? Assume a 9.6 μs inter-frame gap and maximum frame size.
(b) How long will the file transfer take if the LLC uses an acknowledged datagram service?
(c) If other devices are now using the network such that each frame is subject to an average of 4 collisions, estimate the time taken to transfer the 1 Mbyte file using the unacknowledged LLC service.

2. Estimate the time taken to transmit 1 Mbyte of data using full duplex transmission over High Speed and Gigabit Ethernet for frame sizes of 248 and 1500 octet payloads.
3. Determine the efficiency of the transmission process associated with half duplex Gigabit Ethernet when transmitting minimum size frames, with and without frame-bursting.
4. For both 100Base and 1000Base CSMA/CD networks, estimate how long it will take for 1 Mbyte of data to be transferred via a contention-free half duplex configuration when frame payloads are 248 and 1500 octets.
5. A MAN based upon the FDDI standard consists of 50 nodes, 4km apart, connected by 100 Mbit/s ring. Each node generates an average of 100 kbit/s data traffic, while half of the nodes generates 256 kbit/s of voice traffic and the other half generates 2.5 Mbit/s video traffic which has a real-time delay constraint such that successive frames must not take longer than 12 ms to traverse the MAN.

(a) If it takes the MAC token 600 micro-seconds to circulate the network under no-load connections, estimate the actual and maximum possible utilisation for this network under these conditions.
(b) What options are available to improve the maximum efficiency and what consequences will these have on overall performance?

6

The Channel Perspective

6.1 Introduction

In this chapter, we reach the lowest layer in the communication system, the physical layer. The upper layers have all acted in one way or another to insulate the system from the details of the lower layers, but the physical layer constrains what the communication system is actually able to do. There are two main elements involved here: the physical transmission medium itself, and its capabilities, and line coding and modulation, which is the process of tailoring the raw data presented to the physical layer into a form which can be successfully transported across the channel.

6.2 Channel Capacity

6.2.1 Discrete Memoryless Channel

In Chapter 2, we considered the information content of sources, where a source generates messages. However, in a communication system we are usually more interested in channels. A channel is a conduit for messages – an entity which transports messages from one point in the system to another. Channels may be noiseless – where the output is equivalent to the input (there may be a transform, but there is a one-to-one mapping from the output back to the input), or noisy, where some noise signal is added, and a given output may result from more than one input. In this latter case, we have an input to the channel X and an output Y, and both X and Y are random variables (see Figure 6.1). The channel is normally connected to a source with a known characteristic so that the probability of the input distribution X is known. We are interested in the transmission of information, where the observer will be at the receiver, so the question is therefore what can we learn about X from observations of Y.

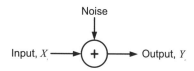

Figure 6.1   Transmission channel with noise

A discrete memoryless channel is the channel equivalent of a discrete memoryless source, i.e., a channel which passes (generates) message symbols with a probability which does not

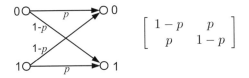

Figure 6.2 Binary symmetric channel

change over time and is independent of previously passed (generated) symbols. The source X can produce symbols from the alphabet $\{x_1, x_2, \ldots, x_J\}$. The received symbols have values $\{y_1, y_2, \ldots, y_K\}$. We further define a set of transition probabilities $p(y_k|x_j) = P(Y = y_k|X = x_j)$, the probability that $Y = y_k$ given that $X = x_j$, for all j and k. We can form these transition probabilities into an array of values, called the channel matrix. This array, along with alphabets A and B, totally define the channel.

$$\begin{bmatrix} p(y_1|x_1) & p(y_2|x_1) & \cdots & p(y_K|x_1) \\ p(y_1|x_2) & p(y_2|x_2) & \cdots & p(y_K|x_2) \\ \vdots & \vdots & \ddots & \vdots \\ p(y_1|x_J) & p(y_2|x_J) & \cdots & p(y_K|x_J) \end{bmatrix}$$

Each row of the array corresponds to a fixed channel input. The addition along these rows must add to 1. Each column corresponds to a fixed channel output.

In many cases the input alphabet and the output alphabet will be the same. A very common channel is the so-called 'binary symmetric channel' with binary input and output alphabets and the transition matrix shown in Figure 6.2.

The binary symmetric channel introduces errors with a probability p.

6.2.2 *Capacity of a Discrete Memoryless Channel*

Recall the concept of mutual information. We defined the information about X conveyed by Y to be $I(X;Y) = H(X) - H(X|Y)$, or the uncertainty of X less the uncertainty remaining about X after Y is known. In the context of a channel, if Y is the received symbol and X is the transmitted symbol, the mutual information is the information that the reception of Y conveys about X, or put another way, the uncertainty about X removed by the reception of Y. The mutual information is the information carried by the channel.

Consider a transition probability $p = 0$. The received symbol is equal to the transmitted symbol, and $H(X|Y) = 0$ (there is no uncertainty about the value of X after Y is received). The mutual information is therefore $H(X)$, the entropy of the source. If $p = 0.5$, then a received bit is equally likely to have been the bit transmitted or its inverse, so no uncertainty is removed regarding the value of the transmitted symbol. $H(X|Y) = H(X)$. The mutual information is now zero.

More formally, the entropy of X given a particular value of Y $(= y_k)$ is

$$H(X|Y = y_k) = -\sum_j P(X = x_j|Y = y_k) \log_2 \left(P(X = x_j|Y = y_k) \right).$$

We can then take this over all values of Y as follows

$$H(X|Y) = \sum_k \left(H(X|Y = y_k) \right) p(y_k) = -\sum_k \sum_j p(x_j, y_k) \log_2 (p(x_j|y_k)).$$

Note that $H(X) = -\sum_j p(x_j) \log_2 p(x_j)$. Since $p(x_j) = \sum_k p(x_j, y_k)$, this gives us that $H(X) = -\sum_j \sum_k p(x_j, y_k) \log_2 p(x_j)$, and so

$$I(X;Y) = H(X) - H(X|Y) = -\sum_j \sum_k p(x_j, y_k) \log_2 \left(p(x_j) p(x_j|y_k) \right).$$

Note that the information carried by the channel depends on the entropy of the information source as well as the transition probabilities. In other words, it depends not only on the channel itself but on the way it is used. This is inconvenient for the study of channels, so we define the maximum information the channel can carry, called the channel capacity, as the maximum average mutual information (between the source and destination) for any single use of the channel, where the maximisation is over all possible source probability distributions $\{p(x_j)\}$ of X. This is measured in bits per channel use. Usually this means per symbol.

Consider the binary symmetric channel $p(y_0|x_1) = p(y_1|x_0) = p$; $p(y_0|x_0) = p(y_1|x_1) = 1 - p$. Channel capacity $C = \max I(X;Y) = 1 + p \log_2 p + (1 - p) \log_2(1 - p)$.

Note that $p \log_2 p + (1 - p) \log_2(1 - p)$ is the entropy of a binary source with probabilities p and $1 - p$. The maximum entropy of a binary source is 1 bit per symbol, so the capacity of a binary symmetric channel can be thought of as 1 (entropy of source) $- H(p)$ (entropy of binary source with probability $p : 1 - p$). If the channel is noise free, $p = 0$, and the channel capacity is 1 bit per symbol, and if $p = \frac{1}{2}$, the channel capacity is 0.

6.2.3 Channel Coding Theorem

What the above shows is that even if a channel introduces errors, it is usually possible to transmit some information through the channel. The amount of information is given by the channel capacity. Shannon proved that if a discrete memoryless source with entropy $H(A)$ producing symbols every T_s seconds transmits over a discrete memoryless channel with capacity C used every T_c seconds, it is possible to construct a coding scheme with an arbitrary small probability of error if $(H(A)/T_s) \leq (C/T_c)$. Further, if $(H(A)/T_s) > (C/T_c)$, the information could not be transmitted with an arbitrary small probability of error. If applied to a binary symmetric channel (BSC), $(1/T_s) \leq (C/T_c)$. $T_c/T_s = r$, the rate of the code scheme. Put another way, if $r \leq C$, it is possible to construct a working code. C depends on p, the transition probability.

6.2.4 Analogue Channels – The Channel Capacity Theorem

The above discussion has applied to discrete sources and channels. In many cases in a digital transmission system we can use these discrete channels, but the real world is analogue, and analogue transmission needs a different approach. If we extend the principle of entropy to analogue signals by expressing the signal in terms of infinitely small discrete divisions, we end up with infinite entropy, as we are totally uncertain in which division to place the analogue signal value. It is possible to remove this factor and use *differential entropy* and use this, but the proofs are complex. A better solution is to return to the original method of Shannon.

Recall Hartley's definition of information $I = \log_2 s$, where s is the number of possible messages. Using an analogue signal, with a bandwidth of B, we can have $2B$ symbols per second, and the channel capacity, $C = 2B \log_2 s$ bits (Hartley's Law). Note that (a) we can increase capacity by increasing the number of symbol levels but (b) Hartley's Law does not tell us the maximum number of symbol levels we can have, and thus cannot give an absolute figure for channel capacity. For this we must consider channel errors, power level and noise.

Shannon's law does this and we proceed as follows. Consider the locus of points within a certain distance from a point in space. The distance is the root of the sum of the square of each dimension (x and y for 2 dimensions, x, y and z for 3 dimensions). This locus forms a circle in 2 dimensions, a sphere in 3 dimensions, and a hypersphere in more than 3 dimensions. If we consider n dimensional space, with $n \gg 1$, an interesting thing happens; most of the 'volume' of these hyperspheres occurs near their edge. For 2 dimensions, only 2% of the area of a circle lies in the outermost 1% (by distance), whereas for 100 dimensions, 99.996% of the volume lies in this outer 1%. Let a source transmit over a channel with bandwidth B every T seconds. If the symbols are x_1, x_2, etc, then the total energy is proportional to $E = x_1^2 + x_2^2 + \ldots + x_n^2$. This is identical to the equation for (distance)^2 from a point to the origin in hyperspace. If the average symbol energy is S, the total message energy is $2BTS$ joules. As $n \to \infty$, these points will tend towards the surface of a hypersphere of radius $\sqrt{2BTS}$. Each symbol may be corrupted by noise, and a similar argument can be used to represent the signal + noise points as hyperspheres of radius $\sqrt{2BTN}$ centred on each message point, where N is the noise power. Due to the surface-hugging nature of the system, we now have a combined hypersphere of radius $\sqrt{2BT(S+N)}$. If the channel is within its capacity, these hyperspheres may not overlap, so capacity will be the number of hyperspheres which can fit into the large hypersphere. The volume of a hypersphere with n dimensions is ar^n for some constant a (a is π for 2 dimensions, $\frac{4}{3}\pi$ for 3 dimensions, etc.). We have $2BT$ dimensions, so the volume of the large hypersphere is $a\sqrt{2BT(S+N)}^{2BT}$, and the small ones is $a\sqrt{2BTN}^{2BT}$. The ratio is the maximum number of distinguishable messages which can be sent, which is $(\frac{S+N}{N})^{BT}$. Channel capacity is therefore

$$\log_2\left(\left(\frac{S+N}{N}\right)^{BT}\right) = BT\log_2\left(1 + \frac{S}{N}\right)$$

T is the duration of each symbol, and although S and N correspond to energies, they both apply to the same time interval and so equal the power ratio. This allows us to state Shannon's Channel Capacity Theorem as

$$C = B\log_2\left(1 + \frac{S}{N}\right) \text{ bit/s.}$$

6.3   Transmission Media

The transmission media is the material that forms the link between the transmitter and the receiver.

6.3.1   Wire

The simplest form of transmission material is a wire. Wires are almost always used today with an insulating cover. Simple wires without any form of shielding are used within equipment and for links over short distances (< 1 metre). If more signals have to be carried, multicore wire is used, where an overall cover contains many independent insulated wires. Multicore wire can be made with the individual cores lying side by side to form a ribbon cable. Ribbon cables can be used with special insulation displacement connectors (IDC) which are squeezed into the ribbon to connect to each of the wires, avoiding the need to solder each wire individually. Wires can be single strand – formed from a single strand of material, or multi-strand, where the wire is formed from a number of smaller strands twisted together. Multi-strand wires are more robust and will withstand movement, whereas single core is slightly cheaper and has less resistance for a given wire size. However, single strand wires can only be used if the wires are not going to be subject to much movement.

There are three issues relating to simple metal wires: resistance, inductance and capacitance. Resistance issues come in two forms – resistance between conductors or to earth, where ideally the value is very high, and resistance within the conductor itself, which should be as low as possible. Modern insulation materials have all but eliminated the problem of insulation between conductors, and most conductors are insulated along their length. Exceptions are power lines. Outdoor uninsulated cables suffer from problems from the rain, as moisture can bridge the insulator between the wire and its support causing significant leakage in wet weather. Old telephone cables were uninsulated, and were fixed to poles using ceramic insulators. These were shaped like bells, so the underside was dry to minimise dampness. Power line insulators have a concertina shape to maximise the surface area, thus increasing the resistance of the surface layer of moisture. Data cables are always now insulated, but water penetration is still a problem for junction boxes, where moisture can cause leakage between terminals.

Resistance along the wire is a function of the resistance of the material used and its cross-sectional area. Copper is a very common material for use in wires, giving a good compromise between low resistance and low cost. Resistance can be lowered by using larger wires with greater cross-sectional area. However, this is more expensive in cost of material for the wire itself. If wires are suspended from posts, larger wires are heavier, requiring more posts, which is expensive. If the wire is buried in trunking, the larger the wire, the fewer can fit in the trunking. For these reasons, rather than increase the size of the wire, it is often preferable to fit repeaters to the wire to regenerate the signal at regular intervals. Repeaters, however, bring their own problems in regard to the provision of power supplies and maintenance.

The second issue for metal conductors is capacitance. When conductors lie side by side, there will be a capacitive coupling between them which will act as if a capacitor had been connected between them. This can be minimised by twisting the wires together to keep them close to each other.

The final problem facing a metal conducting wire is that of inductance. If many wires are carried side by side, a voltage transient in one will induce a current in the neighbouring wires. A simple solution, which is commonly used in ribbon data cables, is to ground the neighbouring wires. On a ribbon cable, this corresponds to grounding every second connector. However, transients can still be induced, and the number of conductors is reduced by half. The best solution is to pair wires so that they form current loops and, for any current flowing from the transmitter to the receiver, there is a corresponding current flowing in the opposite direction from receiver to transmitter. Transients in one will be cancelled out by the corresponding transient in the other direction. This will not counteract induced voltages from other wires, but the effect of these can be minimised by twisting the two paired wires together. The orientation of these wires with the electric field inducing the voltage will change, and the induced voltages will cancel out.

Twisted pair wires are very commonly used for high data rates over relatively short distances (tens of metres), and twisted pair wires are used, for example, in UTP Ethernet with speeds of 100MHz. As the length increases above this the achievable data rate reduces. Different categories of UTP cable are Category 1, for voice or low-speed data < 56 kbits/sec, Category 2, data rates up to 1 Mbits/sec, Category 3, transmission up to 16 MHz, Category 4, transmission up to 20 MHz and Category 5 for transmission up to 100 MHz. Performance can be improved by providing shielding wound round the insulated wires within the outer casing and connected to ground, which allows the distance to be increased to a few kilometres. By using sophisticated line coding and modulation techniques, data rates of megabits per second can still be realised over these sort of distances, and proposals for xSDL allow twisted pair telephone lines to provide high bandwidths to the home.

A metal conducting wire's properties can be further improved for high data rates by using a coaxial cable. This cable consists of a conductor surrounded by an insulator and a solid (highest quality) or braided shield connected to ground. This allows higher data rates to be maintained over longer distances. Thus, 50 Ω baseband cable is used for 10base2 Ethernet, which allows data rates of up to 10Mbit/s over a kilometre, although the relative difficulty of connecting to coaxial cable and the improvements in signal processing and therefore line codes have meant that this type of cabling is now seldom used, as untwisted pair wiring can do the job. At higher rates, coaxial is still the best choice, with 75 Ω broadband cable, which is used, for example, for cable TV, giving throughputs of up to almost 1Gbps and lengths of up to 4km between repeaters. Coaxial cable is often used for metropolitan area communications.

At even higher frequencies, above about 2GHz, it is possible to remove the central conductor completely leaving only the outer shield as a waveguide. These are often used to connect transmitters and receivers to antennas.

6.3.2   Optical Fibre

If we keep on increasing the frequency of the signal we send we eventually reach the optical spectrum. We can send light down an optical fibre – a strand of light-conducting material, glass or some form of plastic, normally covered with an opaque sheath to prevent any extraneous light penetrating the fibre. While cheap optical fibre for short links contains a single light conductor, it is more common for this to be made up of an optical core surrounded by an optical cladding made up of a material with a different refractive index.

Using light avoids many of the problems of metallic wires. There is no induction between fibres and external power sources, and there will be no cross-talk. Illicit tapping of an optical fibre is more difficult. A wire can be tapped by a direct connection or by an inductive coil. It is possible to tap some types of optical fibre, but the effect is much more noticeable than for an electrical wire.

The lack of any electrical signal makes optical fibres safer in environments where sparks could be dangerous, although since the apparatus connected by the fibre are not electrically joined, a separate electrical connection will be required if one is to provide power to the other. In most cases, however, electrical insulation between transmitter and receiver is an advantage. Optical fibres are smaller and lighter than coaxial cable, simplifying installation and meaning that a single coaxial cable can be replaced by a bundle of fibres.

Optical fibre does not suffer from electrical resistive, inductive or capacitive effects, which means that fibres connections can be longer (40-50km between repeaters) and carry higher frequencies. They do suffer from some attenuation, however, from the material which forms the fibre, and have another problem which restricts the highest frequencies carried, that of dispersion.

The speed at which light travels through an optical medium is denoted by the refractive index, η, of the material. Optical fibres work because when light meets a join between two materials with different refractive indexes, η_a and η_b, its velocity changes, and so does the angle of incidence, θ. This is Snell's Law, which gives $\frac{V_a}{V_b} = \frac{\eta_b}{\eta_a} = \frac{\sin \theta_a}{\sin \theta_b}$. If $\eta_a > \eta_b$, $\theta_b > \theta_a$, and there comes a point where $\theta = 90°$. The corresponding θ_a is known as the critical angle, and if the incidence angle is greater than this, the ray of light does not cross into material b but is reflected. An optical fibre is made by surrounding a core with an optical cladding with a lower refractive index.

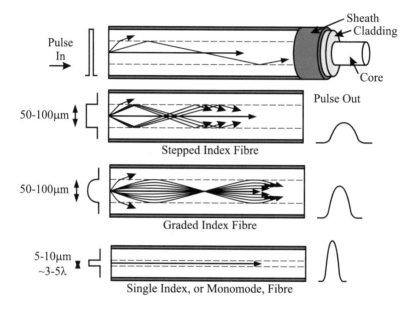

Figure 6.3   Optical fibres

Different types of fibre have different refractive indexes and light conducting properties (see Figure 6.3). Light from a source enters the fibre at a number of different angles, so that while some light may travel straight along the centre of the fibre, much will enter at an angle and reflect back and forth across the core as it travels along the fibre. These different paths are called modes. When a pulse enters the fibre, the light travelling along these reflective paths travels further than those travelling along the core, and takes longer to reach the destination. This causes the pulse to spread out, limiting the maximum transmission rate from the fibre before the pulses overlap to the extent that they cannot be distinguished. A standard multi-mode stepped index fibre, which has a core diameter of about 50 to 100μm, is limited to about 20-200 MHz/km, which limits its usefulness for long distances. Attenuation for these types of fibre is in the range of 10-50 dB/km.

The situation can be improved through the use of graded index fibre, where the refractive index changes over the diameter of the core with lower values near the edge. Not only does this bend the rays of light back towards the centre, but the ray's velocity increases with the lower refractive index, limiting the dispersion of the pulse. Graded index fibres can achieve 200 MHz – 3GHz/km, with attenuation in the order of 7-15 dB/km.

If the core diameter is small enough – in the order of a few wavelengths – only a single wave down the core of the fibre will be propagated, and the only spreading of a pulse will be due to imperfections in the material. Even if the material is perfectly made, there will still be a fundamental limit due to imperfections at a molecular level, and these fibres can be used up to about 3-100 GHz/km. At the wavelengths typically used – 1.3mm or 1.5mm, which have low attenuation through glass – scattering loss due to these effects is about 0.15dB/km, and figures near this have been achieved in the laboratory, although commercial cables typically have losses about 1dB/km.

Single mode fibre is the most expensive type, but is also the most frequently used type for communications links. The cost of the fibre itself is a fairly small part of the cost of installation, and so the higher capacities and reduced need for repeaters possible with single mode fibres more than outweigh their additional cost.

The source of light for the fibre is usually either a Light Emitting Diode (LED) or Injection Laser Diode. LEDs are cheaper and last longer than Injection Laser Diodes, but they do not produce a particularly narrow spectrum of light. A typical LED producing light with a wavelength of about 800nm will actually produce light with wavelengths varying over a range of about 50nm. The different wavelengths will travel at different velocities, causing dispersion in a similar manner to multi-mode fibre, and with a wavelength variation of 50nm, this spreading is about 5ns per kilometre, limiting the transmission rate to about 50 to 100MHz/km. This means that despite the increased cost and greater level of complexity their control requires, Injection Laser Diodes are used for high bandwidth communication links.

Frequency division multiplexing can be used on fibre optics in the same way as other media, only in this case it is termed Wavelength Division Multiplexing (WDM). Injection Laser Diodes are used because of the narrow spectrum of the light they produce, and tens of different wavelengths can be used on a single fibre, each capable of carrying a transmission rate in the order of a single fibre. WDM can be applied to a fibre after it has been installed, allowing retrospective capacity enhancements to be applied to a network.

6.3.3   Radio

Radio is the classic shared medium. Being wireless, and able to travel large distances, it is the obvious transmission medium for mobile terminals. The problem is that there is only one radio spectrum and everyone has to share it. As more and more services are developed, the pressure on the spectrum increases, and higher and higher frequencies are used. Unfortunately, these high frequencies do not propagate well, so range is restricted.

6.3.3.1   Broadcast Radio

Broadcast radio is a very effective means of mass distribution of information. Obvious examples are radio and television, where the fact that a radio signal is used means that terminals do not need a point of attachment for the information feed and this means that services can be deployed with relatively little infrastructure.

This lack of infrastructure, rather than any potential mobility benefits through having a wireless connection, is the main reason that these services used radio in the first place.

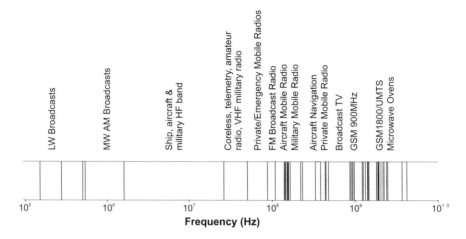

Figure 6.4   Radio frequency bands

However, the lack of available radio spectrum for other services, such as cellular radio systems, which require a wireless link for movement, is increasing the cost of the radio spectrum used by broadcasters. In particular, it means that new spectrum for broadcast services is very hard to come by. This is pushing a move towards digital services rather than analogue, as more services can be fitted in to a given spectrum, and additional 'value added' services can be provided.

Low bandwidth data traffic has also been carried by broadcast services but point-to-point links are a much more efficient use of the spectrum.

Broadcast radio is losing ground in some fixed applications, like television, where for non-mobile equipment a coaxial cable or optical fibre connection can provide a much higher bandwidth for a given user. Unfortunately, the infrastructure costs of this approach are high, and so it is not economic for isolated installations. Satellite systems can provide an alternative in these instances, allowing broadcast radio spectrum to be freed up.

6.3.3.2   Point-to-Point Radio between Fixed Terminals

Microwave Links The main advantage of a point-to-point radio link is that information can be sent between two points without having any infrastructure between them. Such infrastructure may be difficult to provide if the terrain is mountainous, for example, but even over relatively benign country radio has the advantage of not requiring a wayleave to be obtained over the intervening properties, a cable to be laid, etc. While radio amateurs and some low bit data transmission uses lower frequencies, the majority of point-to-point links use high frequency microwave links. The high frequencies used (typically 4, 6 and 11GHz) mean that the wavelength is short and allow the radio beam to be focused in a horn or dish antenna, so that the power can be directed and the signal travel further. The high frequency also allows high transmitted bit rates of about 100Mb/s, but requires a line of sight path between the transmitter and receiver, so the antenna are usually mounted on high towers, with repeaters every 40 to 50 km. In particular for frequencies about 10GHz, the signal can be attenuated by rain and snow.

Wireless Local Loop The lack of infrastructure requirements is important over shorter ranges as well. In a town or city, a firm providing communications to homes or offices would have to go to considerable expense digging up streets to install cables. The alternative is a radio link to an aerial on the premises, which is termed a wireless local loop (WLL), and provides a very low cost means of entering the market. It is also popular in countries with an underdeveloped good communications infrastructure. Current WLL schemes use similar radio technology to cordless phones, but there are proposals to use WLL to provide high bit rate Internet access to homes.

Short range wire replacement Even at very short range, lack of infrastructure in the form of wires has advantages, and one of the problems with the proliferation of portable electronic equipment like computers, PDAs, mobile phones, electronic cameras etc., is the exchange of information between them. Wires are cumbersome and miniature plugs and sockets wear out with use. Infra red held out the promise of wireless interconnection, but only with line of sight. Various radio-based systems have been proposed to avoid this restriction. One of these is Bluetooth, and it can send information at up 384kbit/s using the unlicensed frequency band at 2.4GHz at ranges up to about 10m. The same frequency band is used for Wireless Local Area Networks (WLAN), which use more complex (and therefore more expensive) equipment to transmit at several megabits per second for linking computers. WLAN has an advantage over conventional wired LANs because people can move about an office and remain connected. Some companies are now providing WLANs in public areas such as airport lounges.

6.3.3.3 Mobile Radio

Mobile radio systems are specifically designed to serve terminals which are mobile. While the communications link is point to point, this mobility means that a broadcast signal has to be used, and this will cause interference to other users unless steps are taken to separate out the transmissions.

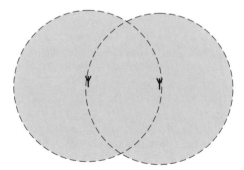

Figure 6.5   Interference caused by mobile radio transmissions

There are several ways of achieving this. The principal method is to divide up the area being served (known as the 'coverage area') into a number of 'cells', individual areas served by one transmitter. Transmitters in different cells use different radio frequencies so that they do not interfere with each other. Frequencies are 'reused' after a certain distance. The smaller the distance, the more frequencies are available in each cell, but the larger the interference will be. The group of cells using different frequencies is known as a clustre (see Figure 6.6), and the number of cells in a cluster, which is the number before the pattern repeats, is known as the 'cluster size'. Users within a cell are divided either by being assigned different frequencies to use (FDMA), different times to transmit (TDMA) or different modulation codes (CDMA). These modulation techniques are discussed in more detail in Section 6.5.

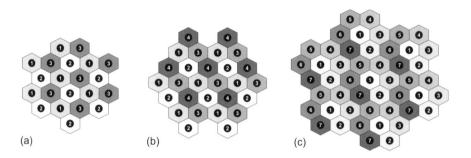

(a)   (b)   (c)

Figure 6.6   Clusters of (a) 3, (b) 4 and (c) 7 cells

The movement inherent in mobile radio systems causes additional problems. The radio signal, being broadcast over a wide area, will be received by the terminal via different paths with slightly different lengths. This means that the components of the transmitted signal will vary in phase leading to constructive and destructive interference. This means that the signal strength at the receiver will change rapidly over a wide range in a process known as 'fast fading'. Fast fading also affects fixed radio systems if objects move near the transmitter or receiver, but to a smaller extent.

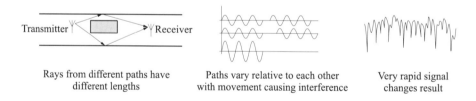

Rays from different paths have   Paths vary relative to each other   Very rapid signal
different lengths   with movement causing interference   changes result

Figure 6.7   Fast fading in a mobile communication system

As the mobile terminal moves, it may move behind obstructions, limiting the signal received. This process is known as slow fading. Finally, the signal strength received will reduce with distance due to propagation losses.

Mobile radio systems have to have complex modulation and receiving circuitry to combat these problems. Often the power used to transmit will be varied in response to the level actually detected by the receiver. This is called 'power control'. The system as a whole also has to have complex resource management functions to ensure that as mobiles move they remain connected to the nearest base station, thus minimising the interference caused to them by other users and vice versa.

6.3.3.4 Satellite

Satellites offer in many ways the ultimate in broadcast radio platforms. Due to their positioning, the signal is largely unaffected by obstructions, except for locations near the pole where geostationary satellites are near the horizon. Satellites are usually used for high bandwidth connections since their high cost means that they have to relay a lot of data to be economic.

Satellites are categorised by their orbit. Satellites in a geostationary orbit, 35,784km above the equator, orbit in the same length of time as it takes the Earth to rotate and therefore appear to be stationary in the sky. Such satellites are used for fixed point to point and broadcast communications where high gain dish antennas can be focused on the satellite. Geostationary satellites have three drawbacks which affect some applications. Their high altitude means that it takes radio waves a significant amount of time to reach them, even at the speed of light. This means that there is a delay of about a 300 milliseconds for the round trip, which is noticeable in a voice call. This distance also causes considerable attenuation, meaning higher transmission power has to be used, which is a problem for portable battery powered equipment. Secondly, the geostationary orbit is becoming very crowded with satellites placing restrictions on their positioning and the frequencies that can be used. Large dish antennas are highly focused and so will ignore satellites even a fraction of a degree away. However, smaller dishes used to receive direct to home broadcast signals are not as directional. The beam width depends on the dish size and the transmission frequency, but a typical 60cm diameter dish used for broadcast TV at 11GHz has a beam width of about 3°, meaning that direct to home broadcast satellites have to be at least that far apart. Finally, since geostationary satellites orbit over the equator, they appear very low in the sky near the poles, and this may cause problems.

For these reasons, low earth orbit satellites are often used for communications. These have lower attenuation and so can be used with mobile phones without high gain dish antennas, and they also have much smaller delays. A smaller area is covered by each satellite, so more are required, but this means that a higher capacity is available.

The life of a satellite is usually constrained by the amount of fuel it has, since they tend to drift in orbit and have to be kept on station. Low earth orbit satellites face greater problems from orbit decay being nearer the atmosphere and have to be replaced every few years.

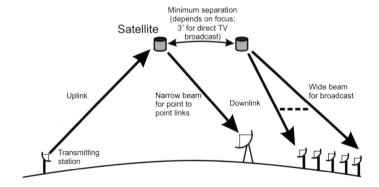

Figure 6.8   Satellite communication systems

6.3.3.5   Relay Aircraft

High capacity mobile radio systems require a good radio channel. One difficult with using ground-based base stations is that the transmitted signal reflects off obstacles and is scattered. This reduces the maximum transmission rate. Satellites offer line of sight radio links but are expensive and suffer from delays due to the distance. A compromise is to use high altitude aircraft or airships to relay radio signals and provide mobile phone coverage. Various consortia have put forward proposals to provide data services either using manned or unmanned aircraft which would stay on station for tens of hours.

6.3.4   Other Media

6.3.4.1   Infra Red

Infra red provides a cheap wireless solution for short-range line of sight links. The fact that the link is line of sight limits interference, but cheap light emitting diode devices are not particularly directional and are not suitable for outdoor use due to the sun. Very short-range links (tens of centimetres) can operate at speeds of a few megabits per second.

  If infra-red lasers are used, much longer outdoor links are possible if the equipment is accurately aligned. Speeds of up to 155 Mbit/s and distances of several kilometres can be achieved allowing organisations to link sub-networks in different buildings, for example, without disturbing the land in between.

6.3.4.2   Inductive Loops

Over very short ranges (up to a few centimetres), inductive loops can be used to transfer information. An inductive loop system involves the use of two coils, and when a current is exerted in one coil it induces a current in the other coil. The voltage induced in the receiving coil is often used to power the receiver, which can be designed without batteries if it only needs to operate in the presence of the transmitting coil. An application is smart cards for payment or access, which can be laid on the transmitter to be interrogated or to undertake a transaction, with any data on the card being stored in non-volatile memory so that can be kept without power. As well as avoiding the need for connection contacts which may wear out, inductive loops can usually work with the receiver some distance from the transmitter, allowing a smart card used for access control to be checked while still in the user's pocket, for example.

6.3.4.3   Power Lines

Various systems have been developed for using power lines to transmit data, either to the home or to electrically driven devices such as trains. The major advantage of this technique is that it saves the provision of additional infrastructure, but there are difficulties in isolation, both of the communications equipment from the power supply, and of different power supplies so that the power grid can be used for several communications links.

6.4   Line Coding

The transmission medium which is used normally cannot accept transmitted symbols in their 'natural' form. For example, a telephone circuit often has isolation coils inserted in it to provide electrical isolation between the two ends of the circuit. This prevents equipment at one end of the circuit drawing power from the equipment at the other if the potential of the earths are different. Such coils are also used for other purposes, to protect against damage from lightning strikes for pole routes, for example.

Line codes put symbols to be transmitted into a form which is suitable for the transmission medium used.

Signal conditioning for transmission has two main forms:

- **Baseband** (normally simply called 'line coding'), where the coded symbols are transmitted directly.
- **Modulated**, where a carrier waveform is manipulated to send the signal.

Line codes have the following requirements

- **Timing**: the line code should produce a waveform which is easy to synchronise so that timing problems do not occur.
- **DC content**: the line code should note have a DC component to allow transmission through ac coupled media.
- **Power Spectrum**: the power spectrum of the signal should be as small as possible to enhance the transmission efficiency.
- **Performance monitoring**: it is desirable to be able to detect if errors have occurred.
- **Low error probability**: the code should have as low a possibility of errors as possible.
- **Transparency**: the code should work whatever pattern of symbols is being transmitted.
- **Complexity**: the line coding and decoding should be simple to reduce the cost of equipment.
- **Uniqueness**: the decoding process should be able to identify the original data.

There are two basic forms of line codes: level codes and transition codes.

- **Level codes**: the information is carried in the voltage or current level of the signal. Two forms:
  - Non return to zero (NRZ) – pulse level maintained throughout the symbol.
  - Return to zero (RZ) – the level returns to zero at the end of each symbol.
- **Transition codes**: the information is carried in the change in level of the voltage or current of the signal.

6.4.1   Binary Line Codes

6.4.1.1   Unipolar Non-Return to Zero and Return to Zero

Unipolar non-return to zero is the simplest possible line coding scheme. A signal (i.e. a voltage or a current) is sent in order to send a 1, and nothing is sent for a 0 (see Figure 6.9). There are two problems with this approach, the first being the fact that a run of 0s or 1s will result in no change of the signal, making sychronisation difficult. The problem can be solved in the case of runs of 1s by sending a pulse for a 1 rather than a constant value, so that the signal returns to zero. This is called unipolar return-to-zero. However, synchronisation of 0s is still an issue. Return to zero also helps combat the other problem, a dc offset, by ensuring that the output changes when there is a 1 rather than simply staying the same, which could mean that it decays away during a run of 1s in a channel which does not transmit dc.

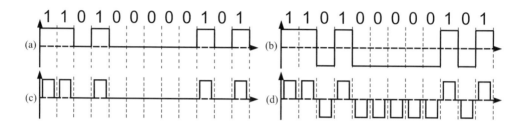

Figure 6.9   (a) Unipolar Non-Return to Zero, (b) Bipolar Non-Return to Zero, (c) Unipolar Return to Zero and (d) Bipolar Return to Zero

A variation on unipolar NRZ is NRZI – Non-Return to Zero, Invert to one – which represents a 0 by the same signal as given in the previous symbol period, and a 1 by a transition. In some senses NRZI is a unipolar version of Bipolar Mark Inversion (see below), but the unipolar nature means that it has a dc component. NRZI is better for clocking purposes than NRZ, since only runs of 0s result in no transitions. In NRZ, runs of either 0 or 1 result in no transitions.

6.4.1.2   Bipolar Non-Return to Zero and Return to Zero

The dc offset inherent in unipolar signalling can be avoided by using –V for the symbol for 0 rather than just 0V. Again, return to zero can be used to give a synchronisation point within each symbol, but at the expense of higher bandwidth requirements. Since zeros now have an explicit symbol, rather than just a lack of signal, synchronisation is still present in a long run of 0s. There is still a problem, however, in that the code is only dc free when there are equal numbers of 0s and 1s being transmitted, and even when this is the case over a long term, in the short term, a dc offset may be present causing transmission problems.

6.4.1.3   Bipolar Alternate Mark Inversion

In this system, 0s are sent by 0V and 1s are represented by alternate highs and lows (see Figure 6.10). This is simple to encode, requiring only a flip flop to record whether the 1 should be represented by a high or a low. The fact that the 1s alternate mean that the code is DC free, but it does have the disadvantage of poor synchronisation, since if there are long runs of zero bits nothing will be transmitted.

6.4.1.4   Manchester Coding

In order to solve the synchronisation problem, Manchester Coding has fixed transition in the centre of each symbol. A 0 is represented by a low followed by high, and a 1 by a high followed by low (see Figure 6.10). The fact that a transition always occurs in the same place

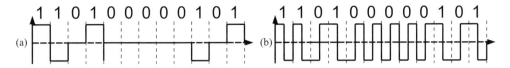

Figure 6.10   (a) Bipolar Alternate Mark Inversion and (b) Manchester coding

allows very good synchronisation, but to ensure this, the symbol must be split in two, which means that with a run of alternate 0 and 1 bits, the number of transitions is twice the bit rate. This is double the rate of line coding schemes we have considered before. Manchester Coding is used as the standard coding method for IEEE 802.3 CSMA/CD LANs.

6.4.1.5   Delay Modulation (Millar Code)

With Delay Modulation, 1s are represented by a mid-symbol change, while 0 has no change (see Figure 6.11). However, to avoid timing promblems caused by runs of 0s, two 0s have a change between them. This means that the maximum frequency is back to $R/2$ (caused by a sequence of 1s), so bandwith use is good, but the fact that transitions can occur either within a burst or at its ends makes the code more complex to detect. A half-symbol phase shift could also result in falsely recorded bits.

6.4.1.6   Coded Mark Inversion

Here 0 is represented by a mid-bit transition, while 1 represented by alternate highs and lows (see Figure 6.11). The maximum frequency is R, but the phase shift problems of Delay Modulation are avoided, although the fact that transitions do not always occur in the same place does increase complexity.

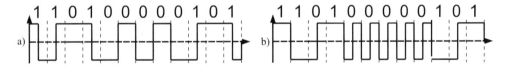

Figure 6.11   (a) Delay modulation (Millar code) and (b) Coded mark inversion

6.4.1.7   Split Phase

This is simpler to detect as each bit has transition in the middle of the symbol, with 1s represented by a change in transition, 0s by the same transition (split phase mark) (see Figure 6.12). Maximum frequency is R. Split phase space is similar, but a change means a 0. Note the similar Manchester coding is sometimes referred to as split phase coding, although Manchester coding is a level code, and this code is a transition code.

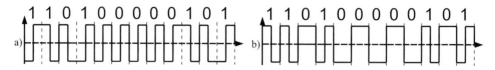

Figure 6.12   (a) Split phase and (b) Biphase line codes

6.4.1.8   Biphase

With biphase, all bits still have a transition at the start for good clock recovery, and 1s also have a transition (for biphase mark) (see Figure 6.12). While the maximum frequency is R, the average frequency is lower, which gives it an advantage over split phase. An alternative (biphase space) has additional transitions on the space (0) rather than the 1.

6.4.1.9   Binary N Zero Substitution

The AMI line code is simple, has good frequency characteristics and no DC component. Its problem is with long runs of zeros, where synchronisation is lost. Fixing this problem by introducing more transitions, as other line codes do, provides good synchronisation, but increases the bandwidth requirements. However, we can fix the problem in another way by replacing the problem runs of zeros with special sequences which include transitions and therefore keep the receiver in synchronisation. These AMI-variant codes are called 'N zero Substitution' codes, where N is the length of the run of zeros the code substitutes. The substituted sequences can be differentiated from normal bit sequences because they violate the AMI rule of switching between +V and –V to represent a '1'.

One of the most popular of these types of line codes is High Density Bipolar 3 (HDB3). Runs of 4 zeros replaced by the special codes 000V or B00V, where B is a normal AMI pulse and V is a 'violation' – a pulse which has the same polarity as the last pulse. 000V is used if there has been an odd number of standard AMI pulses since the last violation, while B00V is used if there has been an even number of standard AMI pulses since last violation.

The fact that the special codes violate the AMI rule means that two codes are required in order to avoid any chance of a dc offset. If there was only one special code – 000V, then it is easy to see that a long run of zeros would cause a dc offset. If the only special code was B00V, then the sequence 000010000100001... would cause a dc offset. Figure 6.13 (a) shows the AMI sequence, which has no dc offset, but few changes, and so poor synchronisation. Figure 6.13 (b) shows a simple form of HDB3 which only uses one type of violating sequence. As can be seen, a dc offset builds up. Figure 6.13 (c) shows true HDB3, where the two different types of violating sequence are used, limiting dc bias.

6.4.2   Multi-level Codes

Making use of additional levels in the output signal can reduce the rate of transitions of the output, and so the bandwidth required to transmit at a given rate. If a bipolar signal is available, MLT-3 (Multi-Level Transition-3) coding can be used. The signal varies from –1 to 0 to +1, back to 0, and then to –1, before repeating. A 1 is encoded by making a transition to the next signal level in the sequence. A 0 is represented by no change. The maximum rate of change of the signal will occur when a run of ones have to be transmitted, when four input bits will run through the sequence. The maximum frequency is therefore a quarter of the bit rate.

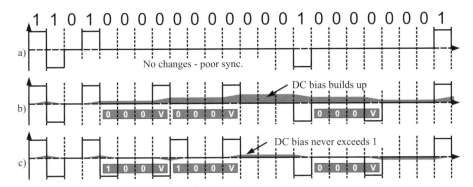

Figure 6.13   Comparison between AMI and HDB3 on a transmitted sequence

The line codes which we have looked at so far have been binary – each symbol transmits a single bit. It is possible to transmit more than one bit per symbol. This means that the baud rate – the rate of transmitted symbols – is different from the bit rate. By transmitting more than one bit per symbol, we can reduce the symbol rate for a given bit rate, and therefore reduce the bandwidth requirements, or we can keep the bandwidth the same and transmit at a higher bit rate. These options are shown in Figure 6.14. In Figure 6.14(a), the bit rate is kept the same, and the frequency over the channel is reduced, shown by the smoother curve. In Figure 6.14(b), the symbol rate, and therefore the transmitted bandwidth, is kept the same, and the bit rate increases (shown by the fact that the same number of bits are transmitted in less time).

The penalty is in terms of bit error rate. The greater number of symbols mean that the difference between the symbols is reduced and the symbols are harder to distinguish. If there is noise in the channel then a smaller amount of noise will be required to change one symbol into another and cause an error.

Figure 6.15 shows a simple multi-level code with four symbol values –3 for 00, –1 for 01, +1 for 11 and +3 for 10. The 00, 01, 11, 10 sequence is called a Gray encoding, which differs from natural binary encoding by arranging for only one bit to change between neighbouring symbols. Since the most likely symbol error is to mistake a symbol for its neighbour, this minimises bit errors.

More complex arrangements are possible. In the 4B3T code (see Figure 6.16), 4 bits are transmitted over a sequence of three tertiary symbols (symbols that can have three values +V (⇑), 0 (–), and –V (⇓). Some data sequences are mapped on to symbol sequences which have no DC component, but some are mapped on to those which do (like ⇓⇑⇑, which has a result of 1). For each of these cases, two possible sequences are defined. The one that is used is the one that will return the overall DC value to nearer zero. 4B3T can achieve over 2.6 times more throughput than that achieved by Manchester encoding for the same maximum bit rate on the channel. 4B3T is used for broadband IEEE 802.3 CSMA/CD.

6.4.3   Scrambling

Communication systems often transmit very regular sequences of bits, and this regularity can have unfortunate side effects. The problem of synchronisation when a long series of zeros is being transmitted has already been mentioned, but regular patterns of ones and zeros could

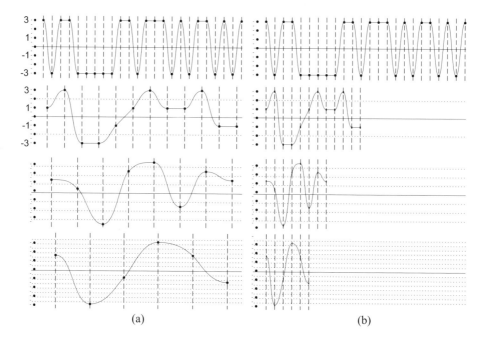

Figure 6.14   Effect of different multi-level codes on transmission time and bandwidth

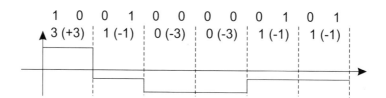

Figure 6.15   2B1Q multi-level code

Data	Output		Data	Output	
0000	⇑–⇓ (0)		1000	⇓⇑⇑ (1)	⇑⇓⇓ (−1)
0001	⇓⇑– (0)		1001	⇑⇑⇓ (1)	⇓⇓⇑ (−1)
0010	-⇓⇑ (0)		1010	⇑⇑⇑ (3)	⇓⇓⇓ (−3)
0011	⇑⇓– (0)		1011	⇑–⇑ (2)	⇓–⇓ (−2)
0100	– –⇑ (1)	– –⇓ (−1)	1100	–⇑⇑ (2)	–⇓⇓ (−2)
0101	–⇑– (1)	–⇓– (−1)	1101	⇑⇑– (2)	⇓⇓– (−2)
0110	⇑– – (1)	⇓– – (−1)	1110	–⇑⇓ (0)	
0111	⇑⇓⇑ (1)	⇓⇑⇓ (−1)	1111	⇓–⇑ (0)	

Figure 6.16   4B3T code

also set up unwanted frequency components or build up DC components in some types of line code. For this reason, a process called scrambling is often used when a 'random' selection of 0s and 1s is added to the transmitted sequence to break up any such patterns (see Figure 6.17). The sequence is actually psuedo-random – it has no apparent pattern but the bits are known so that the receiver can add an identical sequence of bits to recover the original.

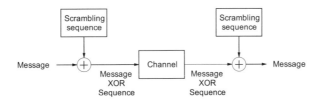

Figure 6.17 Scrambling system overview

An efficient scrambling sequence generator can be constructed using a shift register sequence with feedback designed so that output has good mix of bits. It is started with at least one 1, and if the feedback paths (or taps) are carefully chosen,[1] an n stage shift register will generate a sequence of $2^n - 1$ bits without repeating. For example, a 5 stage shift register with tap polynomial $x^5 + x^2 + 1$ (see Figure 6.18) generates the sequence 0000101011101100011111001010010 and then repeats.

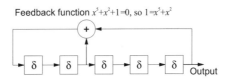

Figure 6.18 Scrambling sequence generator

It is possible to simply add this sequence at the transmitter and receiver, as is done for the LFSR crypographic scheme (see Figure 3.9). However, that raises the issue of initialising the shift register so that it contains an initial '1' and so that both transmitter and receiver start their sequences at the same time. The system can be made self-synchronising by using a slightly different connection as shown in Figure 6.19. As long as no errors occur on the channel, the receiver will automatically synchronise to the transmitter. If errors are likely to occur, resetting the shift register to a known state periodically may be required to stop errors propagating by corrupting the scrambling sequence generator and so adding errors to otherwise correctly received bits.

In mobile radio systems and other systems where a common transmission medium is used, the scrambling sequence is often initialised to a different value for each terminal, so that if the wrong terminal receives the signal it will be descrambled to incorrect values and it will be obvious an error has occurred.

---

[1] 'Carefully chosen' means choosing the tap polynomial to be a primitive polynomial, a polynomial which cannot be factorised and which divides $2^k + 1$ for $k = 2^n - 1$, but which does not divide any $2^k + 1$ for $k < 2^n - 1$, where n is the degree of the polynomial. $x^5 + x^2 + 1$ is such an primitive polynomial.

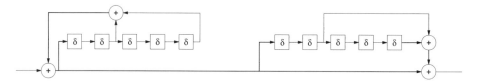

Figure 6.19   Self-synchronising scrambling system

6.4.4   Example Line Code Applications

The RS–232 serial line standard uses Polar NRZ line coding. The coding is simple, and entirely suited to the low transmission rates envisaged in the standard at up to about 9600 symbols/sec. However, it is less appropriate for the higher speeds 'RS–232' interfaces are often used for, where speeds exceed hundreds of kilobytes per second. An RS–232 codeword consists of 5, 6, 7 or 8 data bits, preceded by a start bit of value 0 and trailed by a stop bit of value 1. A parity bit can be defined and the number of stop bits (1, 1.5 or 2) depends on the number of data bits used. If 8 data bits are transmitted, parity cannot be used and there is only a single stop bit.

Bits are inverted for transmission, with a 1 being represented by a voltage below –3 volts and a 0 by a voltage above +3 volts, and bits are transmitted least significant bit first. An ASCII 'A' character, 01000001, would be transmitted as 01000000101. Transmission is asynchronous, so no overall synchronisation signal is required. The normal state of the channel is at the negative voltage level, so the start bit always causes a transition. This causes the receiver to reset its clock and sample bits 0.5, 1.5, 2.5, etc. bit times after this. The stop bit means that the channel is reset to the negative voltage prior to the next start bit. The resetting of the clock on each codeword means that the clock does not have to be too accurate, which keeps the receiver circuitry simple and reduces costs.

Ethernet operates at a much higher speed, and clocking is more important. 10Mbit/s Ethernet uses Manchester coding, which with its mid-symbol transition has very good clocking. The problem is that this transition means that a bit rate of 10Mbit/s requires a channel bandwidth of 10Mbit/s.

MLT-3 is used for Fast Ethernet running at 100Mbit/s, and for FDDI, also at 100Mbit/s, when it is operated over copper wires. Both these schemes use 4B5B encoding in addition to ensure that there are sufficient transitions in the output signal for clocking purposes. The result is considerably more efficient in bandwidth terms than Manchester coding.

The older telephone network standards use simpler, but bandwidth efficient, coding schemes, reflecting times when digital processing was not as fast and cheap as it is now. CCITT G702 trunk standards for E1 (2Mbit/s), E2, and E3 (34Mbit/s) use HDB3 coding, while the 140Mbit/s E4 standard uses Coded Mark Inversion.

6.5   Modulation

6.5.1   Introduction

Many transmission media do not pass DC. A radio or ultrasonic link is unable to transmit DC, and DC on an infra-red link is difficult since it can be mistaken for background light. Even in cases where the transmission media is a plain wire, coils or similar apparatus may be included to prevent the transmission of DC. If data is sent from Glasgow to Edinburgh, for example,

Table 6.1 Example services

Service	Required bandwidth	Actual bandwidth
Speech	3kHz	4kHz (POTS)
Hi-fidelity music	15kHz	15kHz (FM radio) 22kHz (CD)
FM radio station		240kHz
TV (analogue)	4.5MHz	6MHz

but the 'ground' to Glasgow is higher than the 'ground' in Edinburgh, a very large current will flow, and the communication system in Glasgow will start to supply power to the one in Edinburgh. For this reason, isolating coils (or other devices such as opto-isolators) are used to insulate the two ends of the circuit. For telephone lines on pole routes, coils are used to help protect the communication equipment in the event of a lightning strike. Even within the same room, while it is possible to have a common ground, isolating the data line is very useful in case two devices have different voltages or try to drive data line at the same time. The problem with such isolation, either by use of a coil or capacitor, is that the data line is no longer able to transmit DC.

The solution is to transmit an ac sinusoidal signal instead. This is called modulation, and the sinusoidal waveform is called the carrier. We can alter any one of the three properties of the sinusoidal waveform : amplitude (resulting in amplitude modulation), frequency (resulting in frequency modulation) or phase (resulting in phase modulation), or even a combination of these. Another significant benefit is that by using carriers of different frequencies, different information signals can be combined on the same transmission media.

6.5.2   Multiple Access

Transmission media often have to be shared. Table 6.1 gives examples of the bandwidth required for a number of different services. High bandwidth media such as coaxial cable or fibre can carry hundreds of such services, which raises the question of how we can multiplex all the different users onto a single medium. The resource we have to work with is the bandwidth of the medium – the range of frequencies which it can carry – and the time that it is available. Different signals can be differentiated by frequency (Frequency Division Multiple Access), by time (Time Division Multiple Access), or by their modulating signal (Code Division Multiple Access). Combinations of these three are also possible.

6.5.2.1   Frequency Division Multiple Access

Frequency Division Multiple Access (FDMA) is usually used when a fairly large bandwidth is available, such as on radio, fibre and coax. A small range of the available frequencies is allocated to different communication links. There needs to be a guard interval to prevent adjacent channel interference, which means that in practice it is not possible to use the whole range of each frequency band. This guard interval reduces the efficiency of the system, because, as we increase the number of bands and reduce the width of each band, the proportion of the frequency band taken up by the guard intervals increases. The frequency of the modulating signal affects the bandwidth of resulting signal. The band allocated to the signal obviously has to be as wide as or wider than the bandwidth of the signal. This means that as

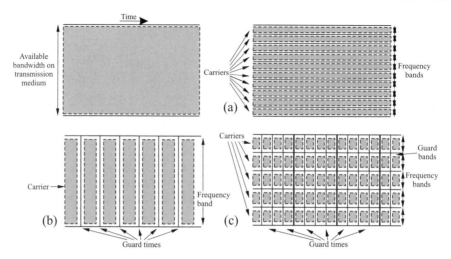

Figure 6.20 Division of resource by (a) frequency; (b) time; and (c) both time and frequency

we reduce the width of each frequency band, we also reduce the amount of information we can transmit in each band. We reach a limit with regards to speech transmission at about 5kHz. Some efficient radio systems use a carrier spacing of 6.25kHz in order to transmit speech signals, but 12.5kHz, or on older systems, 25kHz, is more common.

FDMA is a conceptually simple system, but it has a number of problems. Each link takes up one frequency band, and requires one transmitter and one receiver. In order to have a bi-directions (duplex) link, we need to be able to transmit and receive at the same time and have a transmitter and receiver active at the same time, which requires more complex circuitry. A more significant problem occurs with the transmission of data at multiple rates. If we want to increase the rate at which we transmit data in an FDMA system we have either to increase the width of each frequency band, or we have to use several bands. If we increase the width of each band then those bands which do not require the higher data rates waste resources. Using more than one frequency band to increase the data rate increases complexity. However, if the different users all require the same data rate (speech signals, or analogue TV signals, for example), FDMA offers a simple and reliable solution which does not require the complex synchronisation required with TDMA systems.

6.5.2.2 Time Division Multiple Access (TDMA)

If we are transmitting a digital signal, it is possible for us to store and transmit the signal in bursts rather than continuously. Consider a digital signal which is transmitted at a rate of 1000 bits per second. If this signal is transmitted continuously, we receive one bit every millisecond. However, if we store the received signal and only look at the output occasionally, it is not necessary for the signal to be transmitted continuously.

Consider an observer recording the signal. Rather than recording the bits individually, the recorder works in batches, recording what was received every 10 milliseconds. After this time, we have received 10 bits, each received in one of the preceding milliseconds. But if instead of transmitting one bit every millisecond, we paused the transmission, and then transmitted all these ten bursts in the last millisecond, as far as the observer is concerned, the situation is unchanged and the same 10 bits have been received in the preceding 10 milliseconds (see Figure 6.21).

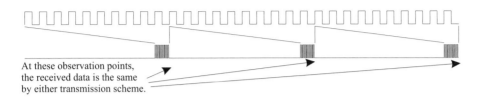

Figure 6.21   Sending information in bursts

We can extend this principle to other time intervals. If we record the signal every 100 milliseconds, we have only to transmit the 100 bits at some point during the 100 milliseconds, but, as far as the observer is concerned, it is unimportant exactly when during the interval the bits are transmitted or at what rate. We could transmit the bits continuously at a rate of 1kbit/s, or we could send them in a 10 millisecond burst at 10kbit/s, or even in a single millisecond at 100kbit/s, or some other combination.

What does change is the delay to the signal. If the signal is sent continuously, there is almost no delay. If we send the signal in a burst every 10 milliseconds, then while the last bit to be sent is sent with almost no delay, the first bit has to wait almost 10 milliseconds to be transmitted (see Figure 6.22). Therefore, the maximum delay is this value. For many information streams, the system must output the original signal at the original rate (as is the case for speech, for example). In such cases, all bits must be delayed by the maximum delay to preserve the overall rate.

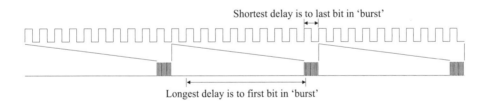

Figure 6.22   Delays in a burst transmission system

Transmitting information in bursts has a number of possibilities. The first, and most important, observation is that it is now possible to multiplex more than one stream of information on to the channel. We can do this by using the space left when the channel is unused to transmit the information of other users in a similar fashion. This is Time Division Multiplexing – where we transmit information on the channel at a faster rate than we need for an individual user and then share out the time between users. Each individual section of information is called a burst, and a sequence of bursts make up a frame, which is repeated continuously. The frame is divided into 'slots', each of which carries a burst. This process, for a TDMA system with four users and therefore four bursts to a frame is shown in Figure 6.23.

As well as sharing resources between different users, we also have the opportunity to incorporate control information in addition to the normal flow of user data by allocating further bursts specifically for such a purpose.

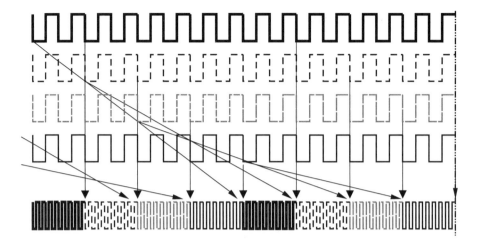

Figure 6.23   TDM transmission system

A major question in the design of a TDM system is the number of bursts which make up a frame and the rate of information transfer on the carrier. For a given length of frame we can increase the transmission rate and either increase the number of slots or the amount of information in each slot, so major options are: the number of bursts which make up a frame, and the rate of information transfer on the carrier. The constraints affecting the choice are the maximum delay which the service can tolerate, as the frame length must be less than this, and the bandwidth of the carrier.

6.5.2.3   Combining FDM and TDM

If we increase the data rate, we increase the bandwidth of the transmitted signal. In a pure TDM system, we will want this to be the entire available bandwidth, but it is sometimes usual to combine TDM and FDM by splitting the bandwidth up into different carriers with FDM and using TDM on each carrier. This is the system used in radio, which the radio spectrum must be shared between different users. For high bandwidth media, a pure TDM system would have to transmit at such a high rate that circuitry would be very complex and expensive, which is another reason why a combined FDMA/TDMA system may be used. FDM systems are conceptually simple and easy to realise up to very high rates (even tens of GHz). TDM has the advantage that it deals with information in packets rather than continuous streams, and packetised data is easier to switch and route. The multiple transceivers of an FDM system are replaced by more complex switching, but there is a problem with getting switches to work fast enough, and this becomes very expensive above a few hundred MHz. FDMA/TDMA combinations often give the best compromise.

6.5.3   Digital Modulation

6.5.3.1   Types of Modulation

In order to have frequency division multiplexing, we need to transmit at frequencies other than those starting at 0Hz. This can be done by modulating a sinusoidal carrier signal of an appropriate frequency. A sinusoidal signal has the form $v_c(t) = A\cos(2\pi f_c t + \phi)$. There are three parameters which can be varied:

- Varying A_c gives *amplitude* modulation (AM) (see Figure 6.24 (a))
- Varying f_c gives *frequency* modulation (FM) (see Figure 6.24 (b))
- Varying ϕ gives *phase* modulation (PM)

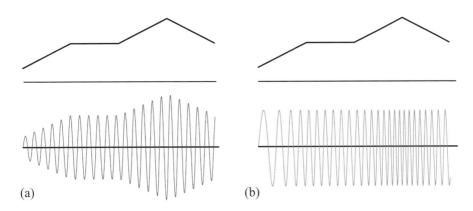

(a)   (b)

Figure 6.24   (a) AM modulation and (b) FM modulation

6.5.3.2   Pulse Shaping

A rectangular pulse such as those used for line coding has instantaneous changes in the time domain which translates to an infinitely wide bandwidth in the frequency domain. The Fourier Transform of a rectangular pulse is a sinc function (see Figure 6.25).

To reduce the transmitted bandwidth to a reasonable level, the signal is filtered. Unfortunately, filtering the signal is equivalent to smoothing it in the time domain and causes symbols to spread in to each other, causing inter-symbol interference (ISI). Raised Cosine Filters, which have a frequency response shaped like a cosine curve raised by the value 1 are often used. A value r, the roll-off factor, is defined, and the response of the filter is flat with a gain of 1 between 0 and $(1-r)f_c$. Between $(1-r)f_c$ and $(1+r)f_c$ the filter response follows a cosine curve, and above $(1+r)f_c$, the response is 0.

If $r = 0$, the filter is a perfect low pass filter. This would mean that no excess bandwidth was required, as all extraneous signals would be filtered out, but unfortunately such a filter cannot be built. $r = 1$, on the other hand, is easy to realise, but has an infinite bandwidth. For bandwidth B, the maximum rate of symbol transmission without ISI is $R = \frac{2B}{1-r}$. Values of r about 1/3 are quite common in practice.

The effect of filtering on the time domain signal is shown in Figure 6.27. When transitions occur between symbols, where discontinuities would normally cause high frequency components, the amplitude of the signal is reduced to a low level. This means that the high frequency components have a very low signal strength.

Time domain   Frequency domain

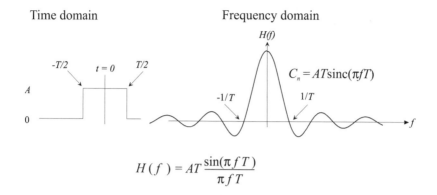

$$H(f) = AT \frac{\sin(\pi f T)}{\pi f T}$$

Figure 6.25   Fourier transform of a rectangular pulse

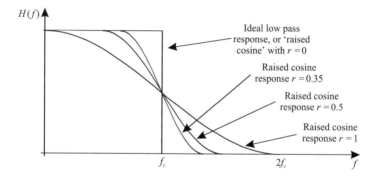

Figure 6.26   Raised cosine pulses

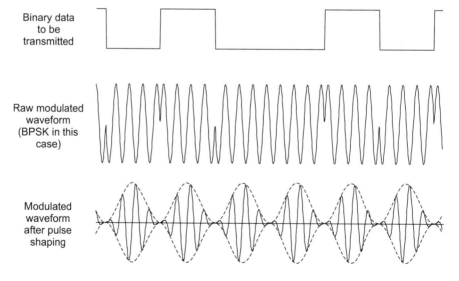

Figure 6.27   Effect of filtering shown in the time domain

6.5.3.3   Amplitude Shift Keying (ASK)

The simplest form of modulation is amplitude modulation, where the amplitude of the carrier is varied with the information signal (see Figure 6.28). If we apply this to digital signals we have Amplitude Shift Keying (ASK). Demodulation of amplitude modulation can be achieved very simply by rectifying the waveform and smoothing it. The transmitted signal is therefore $v_c(t) = h(t)\cos(2\pi f_c t)$, where $h(t) = A_1$ or A_2 depending on the transmitted symbol.

An even simpler form is 'on-off keying' (OOK), where the signal is present to send a 1 or absent to send a 0. OOK is ASK with two amplitudes – 0 and 1 – and can be constructed simply by using a switch which either blocks the carrier or allows it to pass. While OOK is very simple to modulate and demodulate, the fact that there is no signal for a '0' can lead to difficulties with synchronisation, since timing cannot be derived from the carrier.

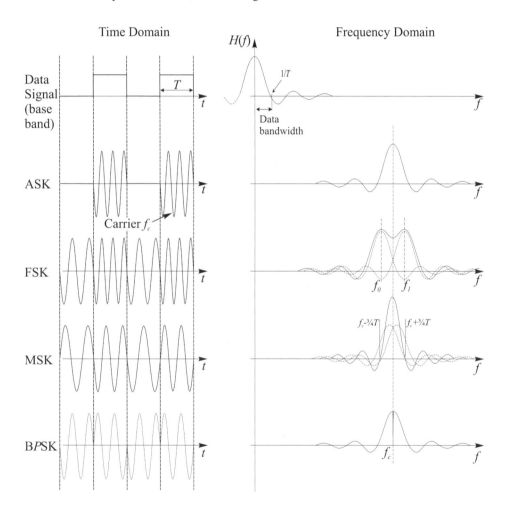

Figure 6.28   Time and frequency characteristics of different modulation schemes

The bandwidth required for ASK is

$$B_{ASK} = (1 + r)R$$

where r is the roll-off factor for the pulse filter and R is bit rate. Depending on pulse shaping therefore, the bandwidth requirement will be between R and $2R$, although a bandwidth as low as R is not practically realisable.

ASK is rarely used today, as its advantages of simple modulation and demodulation are outweighed by its relatively poor performance. ASK, like other forms of amplitude modulation, is not as easy to regenerate as phase or frequency modulated signals. Amplitude modulated signals start out with only two amplitude levels, but the effort of pulse shaping is such that any distortion when amplifying the signal increases the spectrum of the signal, which means that linear amplifiers are required. However, linear amplifiers are less efficient in power usage, and also mean that noise is amplified with the signal. Where only two signal levels are required, phase modulation gives a better performance. However, ASK still finds application is some optical fibre transmissions and for low bit rates transmission on simple equipment (remote controls, for example).

A variant of ASK is m-ary Amplitude Shift Keying. Here there are more than 2 amplitude levels per symbol (m levels). The bandwidth requirement is unchanged in terms of symbols but in terms of bits becomes

$$B_{ASK} = (1 + r)\frac{R}{\log_2 m}$$

where again r is the roll-off factor, R is bit rate, and the factor $\log_2 m$ is included since the symbol rate is $R/\log_2 m$. While this gives the impression of giving higher transmission rate at no additional cost, there is a cost in terms of error performance, since the differences between the amplitude levels is reduced. This means that the receiver must be more sensitive, and limits the electrical noise which can be tolerated on the transmission media.

6.5.3.4   Frequency Shift Keying (FSK)

In Frequency Shift Keying, information is sent by changing the frequency of the carrier signal (see Figure 6.28). Binary Frequency Shift Keying, where two frequencies are used, is most common. However, more than two frequencies can be used, which allows more than one bit of information to be sent per symbol.

The bandwidth required for FSK can be found by considering the signal to be made up of a pair of ASK signals (see Figure 6.28). Each ASK signal has a bandwidth of $B_{ASK} = (1+r)R$ where r is the roll-off factor and R is bit rate. If the spacing between the FSK frequencies is Δf, the resulting bandwidth requirement is

$$B_{FSK} = \Delta f + (1 + r)R$$

FSK has a larger bandwidth than ASK, and is slightly more complex. However, it has good performance, in particular when the number of frequencies is increased (m-ary FSK). The problem is that while this actually improves error performance (for other techniques the opposite is the case), it also increases the required bandwidth, and bandwidth is often a more significant constraint than power (except for very long range communications to spacecraft, for example). MSK (see below) minimises frequency use and is more popular. Early 300 baud modems used FSK, the V21 standard defining two FSK signals (one for each direction) at 1080 ± 100Hz and 1750 ± 100 Hz with bandlimiting of 600Hz.

6.5.3.5 Minimum Shift Keying (MSK)

Since the bandwidth requirement of FSK depends on Δf, the difference between the two frequencies used for the different symbols, an interesting question is how small can Δf be made. The smallest possible frequency difference occurs where the higher frequency has one additional half cycle in the symbol period (see Figure 6.28). Other differences would lead to a discontinuity. This gives that $\Delta f = (f_1 - f_0) = 1/(2T)$, and consequently that

$$B_{FSK} = R/2 + (1+r)R$$

MSK is used as a basis for some mobile communication systems, such as GSM cellular system and the TETRAPOL PMR system where narrow bandwidth is important.

6.5.3.6 Binary Phase Shift Keying (BPSK)

Phase Shift Keying (PSK) uses phase changes to send information. The simplest form is Binary PSK. The carrier is sent to represent a zero, and the carrier shifted by 180° is sent for a one (see Figure 6.28). Since a phase shift of 180° is the same as inverting the carrier, BPSK is equivalent to ASK where the amplitudes are +A and –A.

This equivalence to ASK shows that the bandwidth requirement is the same at

$$B_{PSK} = (1+r)R$$

where r = roll-off factor, R is bit rate. The fact that the carrier amplitudes for the two symbols are the inverse of each other means that on average there is no carrier component in the PSK waveform. While this is efficient in terms that all the transmission power is used for the information signal and none is wasted on the carrier, it does make the receiver circuitry slightly more complex because it has to regenerate the carrier for detection. For this reason, differential techniques are often used.

6.5.3.7 Differential Phase Shift Keying (DPSK)

In a PSK system, a detector is required to detect absolute changes in phase. This can be, in practice, difficult to do and prone to error - in particular any transmitted signal will be subject to random changes in phase. To overcome these difficulties, Differential PSK can be used. In this case, the phase is changed relative to the previous bit transmitted. As shown in Figure 6.29, successive 1s are represented by a phase change whereas transmitting a 0 results in no change. DPSK obviously requires the same bandwidth as a PSK signal which is the same as ASK.

Since the receiver can use the signal received in the last symbol period as a reference, differential modulation schemes are suitable for non-coherent detection. The problem with differential schemes is that the lack of an absolute reference leads to error propagation. If an error does occur, it will affect not only that symbol but subsequent symbols as well.

6.5.3.8 Quadrature Phase Shift Keying (QPSK)

It is possible to use larger numbers of phase changes. If we use 90° shifts rather than 180° we get Quadrature Phase Shift Keying which has four possible values per symbol.

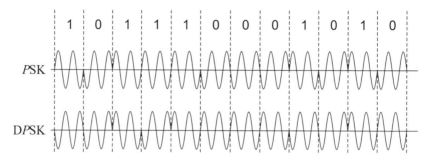

Figure 6.29 Comparison between PSK and differential PSK

QPSK uses the same power as BPSK, but has the same error performance if the decoder can separate the in phase and quadrature components of the signal completely (since the result is then two BPSK signals). However, in practice it is not possible to do this perfectly so there is some performance loss compared to BPSK. However, for a given symbol rate, and therefore signal bandwidth, the bit rate of QPSK is double that of BPSK since one QPSK symbol can encode two bits.

Rather than using standard binary encoding, we can use a Gray Code. A Gray Code ensures that only one bit changes between any two neighbouring points, which will minimise bit errors since a symbol is most likely to be mistaken for a neighbouring point rather than the one 180° out of phase. The Gray Code sequence is 00, 01, 11 and 10, cycling back to 00, rather than the standard binary 00, 01, 10 and 11, cycling back to 00, which has two double bit changes from 01 to 10 and from 11 to 00 (see Figure 6.30).

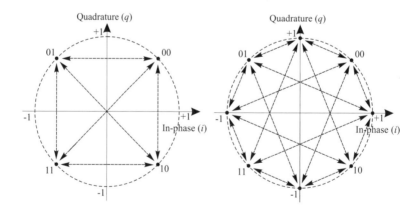

Figure 6.30 QPSK and $\pi/4$–DQPSK symbol constellations

Differential encoding can be used with QPSK for the same reasons as BPSK, with bits being encoded by the change of phase rather than the absolute value. A slight variation is $\pi/4$–DQPSK, where there is also an additional $\pi/4$ phase shift is introduced between each symbol (see Figure 6.30). This additional phase shift means that the transitions between symbols do not pass through the origin, reducing the variation in amplitude. This reduces the bandwidth requirement in combination with pulse shaping.

6.5.3.9 *m*-ary Phase Shift Keying

It is possible to increase the number of symbol points by reducing the phase change between them. However, unlike the transition from BPSK to QPSK, the spacing between the points will reduce as the number of points increase, so although the bit rate will go up for a given symbol rate, the error performance will reduce.

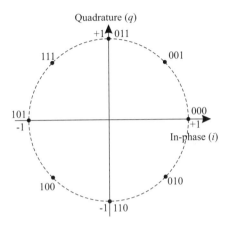

Figure 6.31   Signal constellation for 8PSK

Bandwidth requirement is unchanged in terms of symbols but in terms of bits becomes $B = (1 + r)R/\log_2 m$, where m is the number of symbol values. Therefore the bandwidth requirement of QPSK is $B_{QPSK} = (1 + r)R/2$, and of 8PSK is $B_{8PSK} = (1 + r)R/3$ (see Figure 6.31). In each case where r is the filter roll-off factor and R is bit rate. As m increases, bandwidth requirements decrease, but so does noise immunity.

6.5.3.10   Quadrature Amplitude Modulation (QAM)

If we vary both amplitude and phase components of the carrier we can send more information. The most common form is Quadrature Amplitude Modulation (QAM), which is formed from the addition of two ASK signals 90° out of phase. Depending on the number of different amplitudes of the constituent ASK signals, the signal constellation consists of a rectangular grid of signal points (see Figure 6.32). Other combinations of amplitude and phase are also possible, and form a class of modulation called Amplitude and Phase Keying (APK). QAM is often called 'square QAM' to distinguish it from these other forms.

In its most basic form, square QAM has four distinct symbols, and can therefore encode pairs of input bits. It is generated by adding together two ASK signals with amplitudes of $+/-1/\sqrt{2}$, resulting in a QAM signal with an amplitude of 1. A circuit to do this is shown in Figure 6.33.

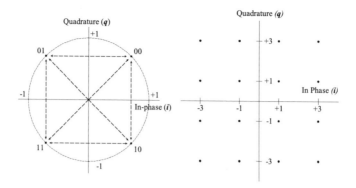

Figure 6.32   Square QAM and square 16-QAM

Di-bit	Phase shift	In phase amplitude i	Quadrature amplitude q
00	$\pi/4$	$+1/\sqrt{2}$	$+1/\sqrt{2}$
01	$3\pi/4$	$-1/\sqrt{2}$	$+1/\sqrt{2}$
11	$-3\pi/4$	$-1/\sqrt{2}$	$-1/\sqrt{2}$
10	$-\pi/4$	$+1/\sqrt{2}$	$-1/\sqrt{2}$

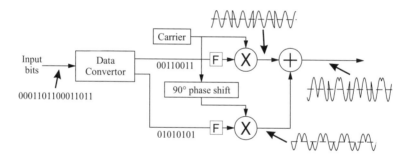

Figure 6.33   QAM modulator

Pure QAM includes sharp transitions at the symbol boundaries as shown in Figure 6.33. This means that practical modulators will include pulse shaping filters at points F prior to multiplying by the carrier.

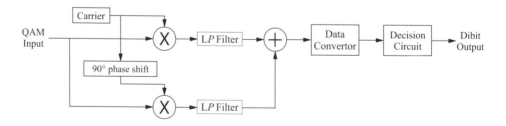

Figure 6.34   QAM demodulator

The amplitude and phase components of square QAM are complex to distinguish, and in some environments this can cause significant difficulties. A radio channel is a good example of this. Other constellations can give advantages. Two common examples are star QAM and circular QAM, as shown in Figure 6.35, although strictly speaking they are APK signals rather than QAM signals.

Star QAM has good average power requirements for a given minimum distance between signal points. Since error rates depend on this minimum distance, star QAM is useful for minimising error rates for a given transmission power. Star QAM also means that signals with a given amplitude are 90° apart in phase, making them easier to distinguish in a channel where phase distortion is more likely than amplitude distortion, such as a telephone channel, and star QAM is used for some modem standards such as V29.

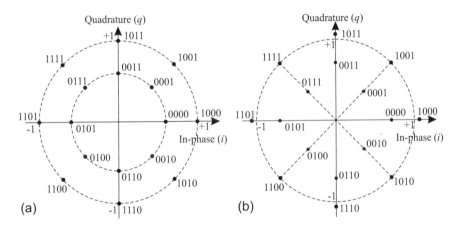

Figure 6.35   APK variants (a) circular QAM and (b) star QAM

Circular QAM is PSK with two or more amplitude levels. The advantage of circular QAM over other types is that differential detection is easier to arrange, which is important as it dispenses with the need to regenerate the carrier.

The bandwidth requirement of QAM is unchanged from ASK in terms of symbols but in terms of bits becomes $B_{ASK} = (1 + r)R/\log_2 m$ where r = roll off factor, R is bit rate. 4-QAM has same bit error performance as $-1/1$ ASK (or BPSK), but m-ary QAM has poorer symbol performance with errors since the signal points are closer. However, it performs better than m-ary PSK, except that 4-QAM and QPSK are equivalent, and therefore have the same performance.

APK is a very efficient modulation system, especially for higher order modulations such as 64QAM or even 256QAM, which is used for cable TV modem systems. However, on channels with both amplitude and phase distortion (such as radio channels, for example), it can be difficult to distinguish changes introduced by the channel from changes intended by the transmitter. This requires complex estimation of the current channel characteristics, which is why QAM schemes are a relatively recent introduction to mobile radio systems.

6.6   Application Example: ADSL

We end this chapter on the channel perspective by looking at Asymmetric Digital Subscriber Line (ADSL). ADSL is an interesting technology in that it shows the use of modulation to extend the capabilities of an existing transmission medium, the twisted pair wire to the home. It illustrates the changing economies of communication systems, where as the cost of processing hardware has fallen, but the cost of installing new transmission media has risen, it becomes more efficient to extend the capabilities of existing media through the use of more complex transmitting and receiving apparatus.

ADSL is a significant element within a group of Digital Subscriber Line (DSL) modem technologies referred to as xDSL. Others within this family include HDSL, SDSL and VDSL. Such techniques are an attempt to provide relatively high speed connectivity between the customer and the communication network that utilises the currently installed copper twisted pair connections that exist within the local loop. Depending upon exact conditions, transmission rates in the region of 10s of Mbit/s are possible. Alternative techniques such as wireless access, optical fibre connectivity and cable modems can also be applied to provide a high speed access but these require the installation of new communication media between the user and the network. Given the sheer volume of current copper connections and the cost of any new cabling installation, there is significant merit in utilising already installed plant. xDSL is used simply by the installation of appropriate modems within the customer's premises and at the local exchange. The deployment of ADSL (for example) is now a political and economic issue rather than just a technical one; the standards for xDSL are significantly advanced as not to be a barrier. It is a matter for the market, the telco and regulators to determine the ultimate widespread deployment of ADSL.

ADSL was standardised in 1995 (ANSI T1.415) and was originally driven by the perceived (but not realised) need/market for Video on Demand (VoD) services. In a VoD service, the transmission capabilities TO the customer/user would be greater than the transmission capabilities FROM the customer, hence the need for asymmetric transmission. Such a characteristic is echoed in today's Internet world, whereby users generate much lower traffic rates than they source – calling a WWW page takes significantly less bandwidth than it takes to download it. It is still thought that there is significant role to be played by ADSL in allowing users to harness the full power of the Internet and remove the 56 kbit/s modem bottleneck.

Twisted pair copper cables, although restricted in a telephony environment to operate in the 0–4 kHz region, are capable of transmitting signals over a bandwidth in excess of 1 MHz. ADSL achieves this through the use of FDM techniques that divide the spectrum into a number of 4 kHz blocks which are used to carry data flows up or down stream between the user and exchange as well as maintaining a dedicated channel between 0–4 kHz to carry voice traffic. The first 25 kHz of spectrum is allocated to carry the 4 kHz voice signal – the unused capacity within this band is used to prevent cross-talk between the voice and data signals. Using the first 4kHz to carry voice enables relatively simple interworking with the PSTN via the use of splitter devices. The signals above 25 kHz are used to convey information between the user and the broadband data network; achieved through the use of ADSL modem pairs.

Data transfer in ADSL is achieved using either Discrete Multitone (DMT) or Carrierless Amplitude and Phase (CAP)/QAM techniques. The former is the standard mechanism and will now be briefly described. The basic concept behind DMT is, in essence, quite simple but which belies its complexity of implementation. The spectrum is divided into a number of individual channels (256) which are each associated with their own carrier signal. Each

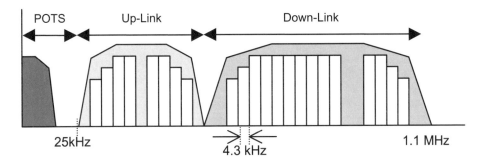

Figure 6.36 ADSL spectrum

carrier is then modulated using a form of QAM enabling a potential capacity in the order of 60 kbit/s. The individual modulated carriers are then simply added together giving a total potential capacity of 15.36 Mbit/s.

The frequency characteristics of each individual channel are not uniform and the information-carrying capabilities of each channel are variable; conditional on the distance and condition of the twisted pair. In general, the frequency response degrades as the carrier frequency increases, which means that the information-carrying capacity is reduced at the higher end of the spectrum. At set-up, the modem determines the quality of each carrier and adjusts the bit-rate accordingly. Two data bands are established; a larger (< 256 channels) downstream band and a small 32 channel upstream band. These bands are either frequency separate bands as shown in Figure 6.36 or the smaller upstream band shares the lower part of the downstream band but is kept separate using echo cancellation. Echo cancellation is a procedure by which, although the two signals are combined, it is possible to retrieve the signal from the remote transmitter since each terminal knows what they themselves are transmitting. In the latter approach, the echo cancellation stage increases the complexity of the modem devices but does offer overall transmission improvement and flexibility.

Although ADSL seems to offer transmission rates above 10 Mbit/s, a number of technical factors limit the rates between 1.5–9 Mbits/sec in the downstream and 16–640 kbit/sec in the upstream. These factors include line attenuation, cable balance, RF interference and cross-talk either at the near or far ends. A combination of interleaving and FEC techniques using Reed Solomon codes are applied to reduce the effects of noise. Both line length and RF signals affect the frequency response and the information-carrying capabilities of any individual set of cables but ADSL is adaptive and can vary bandwidth to suit the prevailing connection and maximise the bandwidth offered.

6.7   Questions on the Channel Perspective

6.7.1   Questions on Channel Capacity

1. A binary symmetric channel with a transition probability of $\frac{1}{4}$ is cascaded with a second channel also with a transition probability of $\frac{1}{4}$ so that the output of the first channel feeds the second channel. Denote the input to the first channel to be X, the output of the first channel and the input to the second to be Y, and the output of the second to be Z. The source at X has a probability of a 0 as $\frac{1}{3}$, and the probability of a 1 as $\frac{2}{3}$.

   (a) Calculate $H(X|Y)$.

(b) Calculate the mutual information between X and Y.

(c) Calculate $H(X|Z)$.

(d) Calculate the mutual information between X and Z.

2. A Binary Erasure Channel has the following form. What is its capacity in terms of p?

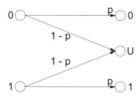

3. A telephone channel has a bandwidth of 3.4kHz. Calculate (a) the channel capacity if the signal to noise ratio is 30dB and (b) the minimum signal to noise ratio required (theoretically) to sustain an information transmission of 4800 bits per second.

7

Putting it all together

7.1 Introduction

In this chapter, we will look at how the design principles covered in the previous chapters have been applied to two example communication systems.

7.2 GSM Mobile Phone System

In our first example, we will look at the design of a digital mobile phone system, sometimes referred to a second generation systems since they followed first generation analogue systems. There are several different standards in use, including cdmaOne^{TM} in the United States and PDC in Japan, each of which chooses to solve the communication problem in different ways, but we will look at the design of the GSM system used throughout Europe and many other parts of the world.

7.2.1 Transmission Media

For a mobile phone system, the transmission medium is the atmosphere. While infra red could work indoors for short ranges, for a mobile system, radio is the only practical choice. Due to the fact that there are many competing uses for radio, use of the spectrum is limited by government regulation. The frequencies available to digital mobile radio systems is a band around 900 MHz (from 890MHz to 915MHz and from 935MHz to 960 MHz).

7.2.2 Line Coding and Modulation

The choice here is either to leave the signal in the baseband and use only line coding, or use modulation as well. Given that the transmission medium requires us to transmit in a narrow band of frequencies, modulation is a necessity.

The use of modulation introduces the choice of modulation type, either amplitude modulation, frequency modulation or phase modulation. Amplitude modulation is unsuitable due to very rapid changes in attenuation present in a radio channel, so it would be difficult to tell if changes in the received signal amplitude are due to the channel or a change in the signal the transmitter is sending. Phase modulation is efficient, but the sharp changes in the output signal can mean a wide frequency band is required unless phase shaping is used. However, pulse shaping causes the output amplitude to change, so linear amplifiers have to be used to avoid distorting the signal. If the exact signal amplitude is unimportant, we can use a nonlinear amplifier without making any material changes to the signal, which is helpful as nonlinear amplifiers are simpler to make and use less power, an important factor for battery-operated equipment. With the improvements in transmitter circuitry over the past ten years, cheaper more efficient linear amplifiers have become available and some radio systems do use phase modulation, but when GSM was being designed this was a risky bet. Therefore, frequency modulation was used for GSM.

Frequency modulation offers the choice of frequency shift keying (FSK) or minimum shift keying (MSK). MSK uses less bandwidth, and so is a better choice for a system which wants to maximise users and therefore revenue. In fact, GSM uses pulse shaping on the binary bits prior to transmission. The pulse shape used is a Gaussian (resulting in Gaussian Minimum Shift Keying, or GMSK), rather than raised cosine, but has the same purpose. Rapid changes of frequency are reduced.

7.2.3 Medium Access Control

The radio channel is a shared medium, therefore some form of media access is required. The requirements in this case are for the system to support a large number of users and provide a flexible service. The two options easily available when GSM was designed were FDMA and TDMA.

With FDMA, each transmitter/receiver pair is given its own frequency band. The system is simple, but has a number of drawbacks in a mobile communication system.

There needs to be a guard band between channels. Speech does not require much information to be transmitted, which implies a small channel. This means that the guard bands will take up a sizeable proportion of the available spectrum, reducing efficiency. Secondly, with FDMA transmissions are continuous. This means that if we want a duplex system (and for speech we obviously require this), each terminal will have to have a transmitter and a receiver which operate at the same time. The problem is slightly more complex, because if a mobile is to be able to sense other transmitters, so it can tell when it is moving out of a cell, for example, then a third unit would be required. This is expensive.

Since GSM is a digital system, TDMA is an alternative. However, if TDMA alone was used, there would be the problem that all users would be on the same carrier frequency, and the only way to limit interference would be to ensure that the corresponding slots in neighbouring cells were not used. This would require base stations to be very tightly aligned in time, and resource management to work over the whole system. Therefore, a combination of FDMA and TDMA is used.

This leads to the question of how to arrange the split. Large carriers reduce the overhead in terms of guard bands. However, this reduces the flexibility to arrange carriers between cells to control interference. The system uses FDMA to split the spectrum into 200kHz carriers, and then TDMA is used to split the carrier into frames of eight slots, each one of which can carry a voice service (see Figure 7.1).

The GSM burst contains two blocks of data separated by a training sequence. The training sequence is a known series of bits which allows the receiver to estimate the channel conditions. The burst shown in Figure 7.1 is the normal burst for speech transmission. There are a number of more specialist burst formats for signalling, although all fit within the same 156.25 bit slot structure.

GSM makes an innovative use of the TDM system to allow for effective duplex operation with a half-duplex transceiver. This reduces costs. With the exception of during its assigned

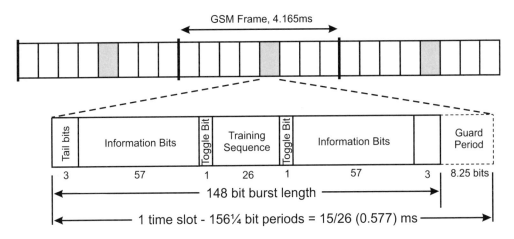

Figure 7.1   GSM normal burst and slot and frame structure

slot, a TDMA system is not transmitting. GSM uses this idle time in two ways. The first is to allow transmission in the other direction. GSM timeslots are numbered such that the uplink slot x is three slot times after downlink slot x. The mobile receives data in its allocated slot x, and then switches to transmit mode to transmit on the uplink slot three timeslots later. This can be compared to a pure FDM system which would require a separate transmitter and receiver to work at the same time in order to achieve full duplex operation (see Figure 7.2). The system also uses the four spare timeslots before it has to receive again to switch to another channel to monitor the signal strength from other base sites to see if it would be more advantageous to switch to a different cell.

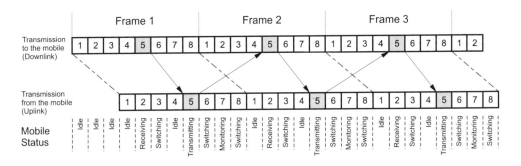

Figure 7.2   GSM achieves duplex communication while only using one direction at any one time

The TDMA structure also allows more flexible service provisioning. By allocating more that one slot in a frame, GSM can accommodate higher data rate services. This is the approach used by High Speed Circuit Switched Data (HSCSD). Alternatively, it is (just) possible to support a speech service using half the normal transmission rate. This is called half-rate speech, and is done be transmitting to a mobile only in every second frame. It requires a more advanced speech coder, but doubles the number of users that can be supported.

7.2.4   Coding

The requirements for speech coding are good, clear speech and a low bit rate. The options are

- Straight PCM
- Adaptive PCM
- Differential PCM
- Transform encoders
- Vector encoders

Since the codec is only going to be used for a single audio source – speech – vector encoders can be used. These require more processing power, but save in the long term because of the reduced transmission rate required. GSM uses a regular pulse excitation/linear predictive coding (RPE/LPC) codec, which encodes speech with 13kbit/s. The half rate speech channel uses a Vector Sum Excited Linear Predictive (VSELP) coder.

A vector codec is very susceptible to errors because of its high rate of compression. Given the very hostile transmission environment (bit error probability up to 1%, compared to 10^{-10} for a good quality fibre), a good error control strategy is required. The options are:

- Error concealment
- Forward Error Correction
- Feedback Error Correction

Feedback error correction cannot be used for speech due to the delays it introduces when used with the TDMA system. By the time a block could be retransmitted, it would no longer be relevant. However, GSM does have a data transmission mode which uses ARQ. A combination of error concealment and Forward Error Correction. is used on the speech channel.

For Forward Error Correction, we can choose between Block codes and Convolutional codes, both with the option of interleaving. Given the high error rates, block codes offering suitable protection would be quite complex. Therefore GSM uses convolutional codes. However, they have quite poor rates, so to reduce the number of bits transmitted, for the speech channel, only the most important bits are protected (for data transmission, all bits are protected). The speech bits are divided into three classes, with the least important being transmitted without any protection.

The other two classes receive some protection, but while some errors can be tolerated, in particular in the least important bits, errors in the most important bits would result in a worse sound than simply concealing the speech frame. Therefore, the most important bits are protected with a cyclic redundancy check (CRC) and then convolutionally encoded. If the CRC of a received block fails, the speech block is concealed by replacing it with silence.

The 456 bits of the encoded speech frame is split into eight blocks of 57 bits each, each of which are transmitted in half a GSM Normal Burst. This gives interleaving over eight frames, which assists the error protection, but does lead to a delay of 40 ms. Each burst therefore carries parts of two speech frames. However, the slot period, at 4.615 ms, is slightly less than the 5 ms which would be needed to transmit these speech blocks. This allows two frames in every 26 to be used for signalling information without disrupting the user data flow.

7.3   Voice over IP (VoIP)

As IP and the Internet become more widespread, the ubiquitous nature of its availability make it a target for transporting one of the most common user services, speech. However, there are

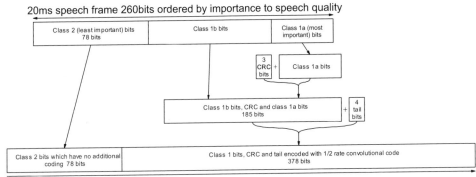

Figure 7.3   Channel coding for the GSM speech coder

Table 7.1   VoIP Speech codec characteristics

Codec	Compression	Sampling Rate	Bit Rate
PCM	PCM	8 kHz sampling	64 kbit/s
G.727	ADPCM	8 kHz sampling	40, 32, 24 or 16kbit/s
GSM	Vector (RPE/LPC)	8 kHz sampling	13 kbit/s
G.729	Vector	8 kHz sampling	8 kbit/s (G.729A)
G.723.3	Vector	8 kHz sampling	6.4 or 5.3 kbit/s
MP3	Various, mainly transform	90 kHz sampling	Various, including 128 and 112 kbit/s

a number of challenges in using IP as the transmission medium for speech, as it is a packet-based system with only a best-effort Quality of Service.

7.3.1   Requirements

Speech can be encoded in a number of codecs resulting in different bit rates and qualities. Standard telephone PCM encoding results in a bit rate of 64kbit/s, while G.723.3 requires only 5.3 kbit/s. Even lower rates are possible – the codec used in the TETRA Private Mobile Radio system uses an ACELP vector codec with a bit rate of 4.567kbit/s, but at these bit rates it begins to get difficult to recognise the speaker (although what they are saying is clear), and this is undesirable for a public telephone application.

Lower bit rates require more complex processing, although all the codecs listed in Table 7.1 give similar performances in terms of Mean Opinion Score, a qualitative measure of speech quality.

Of greater concern than the bit rate requirement, especially given the fact that the underlying transport is packet based, is that of delay. Speech is very sensitive to delay. A delay of less than about 150 milliseconds is not noticeable, but while delays in the order of 150 to 400 milliseconds are tolerable, anything above that is not. This is a very tight requirement, especially if multiple networks are involved. As speech is more tolerant of loss, often it is better to discard heavily delayed packets than to keep them. Network designers attempt to achieve delays of less than about 50 milliseconds for this reason.

Jitter represents the differences in arrival times between data samples or packets, and it also has significant effect upon voice quality. Its effects can be minimised to a certain extent by the use of sequence numbers and time stamps in the voice packets which indicate when the packet was sent, since speech is reasonably tolerant to lost packets and dropping delayed packets may give better results. Adding a playback delay in the receiver, i.e. queuing packets in the receiver for a short time to even out delay, allows delay and loss to be traded, but this increases delay which may be intolerable. Setting too tight a delay constraint will result in any delayed packets being discarded, and this may mean that the loss rate is too high. Adaptive schemes operate by estimating the magnitude and variance of the end-to-end network delay and setting a threshold accordingly. This can be done on the fly by stretching or compressing silence intervals to vary the delay.

7.3.2   Error Control

Since IP is a best effort transport mechanism, packets can be lost in transmission (usually due to network congestion) as well as being deliberately discarded because their transit time exceeded acceptable limits. In real-time applications like speech, recovering from packet loss using packet retransmission does not really make much sense because of the time it would take. This leaves the option or error concealment or forward error correction.

Error concealment involves receivers replacing missing packets either with duplicate packets already received or attempting to interpolate between samples. The latter implies a slight increase in latency as a buffer is required to store samples.

Forward error correction (FEC) requires the transmission of extra data. With VoIP, this extra data can either be in the form of extra (redundant) samples of data or the transmission of an additional lower grade data stream. When packets are lost, the extra data can be used to recover some of the lost stream. As always with FEC, there is a price to be paid in terms of additional transmission bandwidth.

Interleaving can be used to minimise the effect of the loss of a packet. The speech is buffered and the samples reordered before being packetised and transmitted. However, this increases delay, although it has minimal transmission overhead.

7.3.3   Transmission Protocol

We require a link control protocol in order to manage the transmission of the encoded speech packets over the IP network. Two protocols are used. The first is UDP, and on top of UDP sits RTP, or Real Time Protocol

RTP provides a common packet format to carry standard or proprietary multimedia traffic streams. The frame structure of RTP is shown in Figure 7.5. It consists of a minimum of 12 bytes. The first two bits form the version (V) field, identifying the RTP version (currently 2). The next two bits are the P bit, which specifies if padding is added after the payload, which may have been done if the transport protocol or encryption algorithm required a fixed block size, and an X bit, specifying if an extended header is used. The next four bits (CC) specify the number of any CSRC (contributing source) fields. The marker (M) bit allows what the standard defines as 'significant events' to be marked, such as frame boundaries. This is followed by a 7 bit payload type (PT) field identifying the the payload and therefore what the application needs to do to decode it. Examples are given in Figure 7.5. The remainder of the fixed length header consists of a sequence number, a time stamp, and a Synchronisation Source (SSRC) field, which identifies the source. This allows a single device, which would therefore have a single network address, to have multiple sources, which may be different media (audio, video, etc), or different streams of the same type of media. Since the sources may be on different devices, the SSRC identifier is chosen randomly so that the chance of any two sources within an RTP session is very small. However, a mechanism is defined for resolving contention should it arise.

The fixed length RTP header may be followed by up to 15 32 bit CSRC fields. These identify contributing sources. An example of how these may be used is shown in Figure 7.6, where a mixer, which takes RTP packets from two different sources and combines them. On the output from the mixer, the SSRC is the mixer itself, rather than the original sources, but the mixer uses the CSRC fields to identify the original components. Mixers are useful to allow devices with different capabilities to get what they require from a flow.

RTP is supported by another protocol, Real Time Control Protocol (RTCP), which provides additional reports with information on the RTP session. Note that neither UDP nor RTP provide any QoS. This is left to the application.

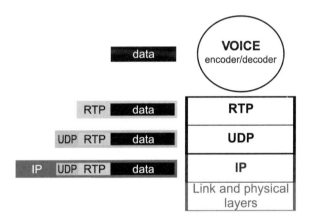

Figure 7.4   VoIP stack

RTCP provides feedback to senders and receivers on such issues are the QoS of the flow, information on the packets (loss, delay, jitter), and end-to-end information on the user, the application or the flow. There are two types of reports, generated either by senders or by receivers to control the flow. For example, information on the proportion of packets lost and the absolute number lost allows the sender (on receipt of the report) to recognise that congestion may be causing receivers not to receive the service packet streams they would expect or need. The sender could then perhaps reduce coding rate to reduce congestion and improve reception. The sender report contains information about when the last RTP packet was generated – both the internal timestamp and the 'real' time. This information will allow the receiver to co-ordinate and synchronise multiple streams, e.g. video and audio streams. If a flow is directed at a number of recipients, there will be RTCP packet flows from each. However, steps are taken to limit the signalling bandwidth by inversely relating the rate at which RTCP reports are generated to the number of receivers.

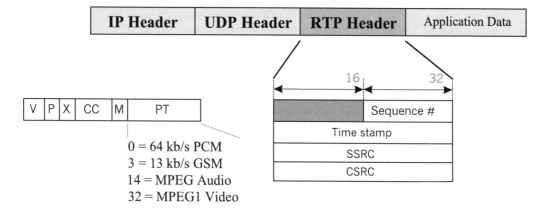

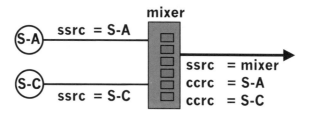

Figure 7.5   VoIP headers

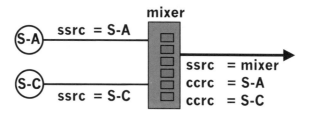

Figure 7.6   Contributed flows after a mixer

While RTCP is provided separately from RTP, the RTP/UDP/IP chain results in a significant overhead in the form of the header. The G.729 codec produces packets of 10 bytes (80 bits every ten milliseconds). The RTP header alone at 12 bytes is larger than this, to which must be added 8 bytes for UDP and 20 bytes for IP (using IPv4), making the header four times the size of the data. Only 20% of the bits actually transmitted are user data. The forthcoming IPv6 increases the header by a further 12 bytes.

The obvious way of reducing this overhead is to transmit larger packets, but larger packets mean a greater sampling time and a higher delay. However, if voice calls are going from the same source to the same destination (which might be the case if VoIP is being used to transport voice calls between an ISP and a telephone network, for example), then the different voice streams can be multiplexed using a slightly expanded RTP header. Another solution is to note that many of the header fields do not change between packets, and so the header can be compressed to remove this redundant information by storing the static information at the source and destination.

7.3.4   Architecture

The above allows two terminals which support VoIP to communicate. However, there is the problem of an overall system architecture which in particular allows communicating parties to set up sessions in the first place.

One possible solution is Session Initiation Protocol (SIP) which defines protocols for identifying terminals and their capabilities over an IP network, either through direct communication with the terminal, or through the use of a Registration Server for location. However, SIP only provides a means of starting a session – it does not support ongoing sessions.

A more general solution which supports session set-up as well as interconnection to the Public Switched Telephony network is the H.323 standard for real-time audio and video conferencing within an Internet environment. The architecture is shown in Figure 7.7. Within H.323, the following elements are defined:

- End Points (Terminals) represent the standalone devices that act as the sources and destinations of the communication session. These endpoints can be PCs and computing devices or can be telephone devices within the PSTN.
- A gateway provides the interface between different network types; normally thought to be the interface devices between the PSTN and the IP network. Usually both data and signalling have to be converted, for example, converting 64 kbit/s PCM streams into RTP packets (and vice versa), as well as converting the signalling between the two domains. The signalling represents the mechanisms by which end-to-end functionality and user services are achieved.
- A gatekeeper provides additional service functionality associated with the communication service. Gatekeepers allow features such as address translation, billing, authorisation, bandwidth management. Gatekeepers can also communicate directly with gateways on behalf of the end points which allows more (flexible) functionality than if direct communication was used. The logical group of endpoints served by single gatekeeper is called a domain.

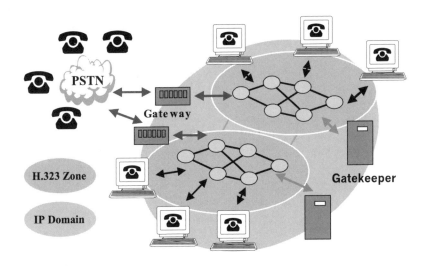

Figure 7.7 VoIP architecture

H.323 is really an umbrella specification that makes use of a range of other protocols and standards to achieve its aims. The protocol stack is divided into two parts; application and control (see Figure 7.8).

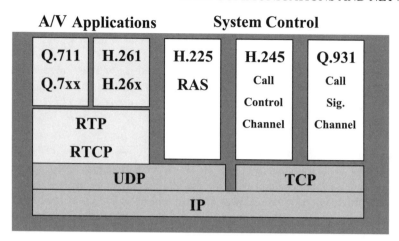

Figure 7.8   User data and control stacks for H.323

The default audio standard is G.711 which represents 64/56 kbits/sec PCM speech. Other G.7xx standards are also supported. Video support is optional but H.26x standards are supported. The application data is encapsulated and sent via RTP and RTCP.

Control is provided at a variety of levels. RAS (Registration/Admission/Status) channel protocol (H.225) allows the endpoints to communicate with the gatekeeper if one is present. Q.931 is responsible for establishing and terminating sessions and provides standard PSTN call functionality between endpoints. H.245 is an out-of-band protocol that is used to establish the media protocols that are used by each H.323 end point.

Another architecture that offers some advantages over H.323 and supports a wider range of networks than IP is MEGACO (Media Gateway Control Protocol). In this network, gateways are supported by a smaller number of gateway controllers, simplifying gateways and reducing their cost.

8

Answers to Exercises

8.1 Answers to Questions on the User Perspective

8.1.1 Answers to Questions on Information Theory

1. *What is the information content (in bits) of the following events:*

(a) *Tossing a coin?*

Information $I = \log_2 s$, where s is the number of events. For tossing a coin, there are two events, so $I = \log_2 2 = 1$ bit.

(b) *Tossing two dice?*

For tossing two dice, there are 2 lots of six possibilities; $I = 2 \times \log_2 6 = 5.170$ bits. The answer can also be obtained by considering the 36 different possible outcomes of tossing two dice, and $I = \log_2 36 = 5.170$ bits.

(c) *Drawing a card from a shuffled deck of cards: considering each card individually; considering each suite separately, considering each card number independently of suite, and considering only the colour?*

In the first case there are 52 events, so $I = \log_2 52 = 5.70$ bits. In the second there are 4 possible outcomes (clubs, spades, hearts, diamonds), so $I = \log_2 4 = 2$ bits. In the third case there are 13 different outcomes, so $I = \log_2 13 = 3.70$ bits. Finally, the cards can be red or black, giving $I = \log_2 2 = 1$ bit of information.

2. *By making reasonable judgements regarding image resolution, estimate the information content of a PAL TV picture.*

The answer depends greatly on the assumptions. Consider each picture element on the screen as one symbol. Each of these symbols has three colours, with, say, 1024 different distinguishable values each. The information content of each picture element is therefore $I = \log_2 1024^3 = 30$ bits. There are 625 lines in a frame, although consider only 600 are visible. Assuming a similar horizontal resolution, there would be 800 elements in each row. The number of elements, which is the number of symbols, is therefore 480000. The information content of each frame is therefore about 14,400,000 bits.

There are 25 such frames per second. However, the actual information in a TV signal, although large, is not this high because this analysis assumes the picture elements can all assume any value. The bandwidth of a TV channel is not high enough for this, and neighbouring elements are highly correlated as a result.

3. *Calculate the entropy of the event of tossing a pair of dice:*

(a) *considering the individual scores of each die (so (1, 6) and (6, 1) are separate events);*

There are 36 possibilities, each with a probability of $\frac{1}{36}$. The entropy is therefore $-36 \times \frac{1}{36} \times \log_2 \frac{1}{36} = 5.170$ bits.

(b) *considering the scores without regard to number (so (1, 6) and (6, 1) are considered to be the same);*

There are 6 possibilities which are the same both ways round, and the remaining 30 options are actually 15 pairs of equivalent results, each of which have two chances (2/36) of occurring. The entropy is therefore

$$-6 \times \frac{1}{36} \times \log_2 \frac{1}{36} - 15 \times \frac{2}{36} \times \log_2 \frac{2}{36} = 0.862 + 3.475 = 4.337 \text{ bits.}$$

(c) *considering only the sum of the scores of the dice.*

There are 11 options as follows

Sum	Occurrences	Number
2	(1, 1)	1
3	(1, 2) (2, 1)	2
4	(1, 3) (2, 2) (3, 1)	3
5	(1, 4) (2, 3) (3, 2) (4, 1)	4
6	(1, 5) (2, 4) (3, 3) (4, 2) (5, 1)	5
7	(1, 6) (2, 5) (3, 4) (4, 3) (5, 2) (6, 1)	6
8	(2, 6) (3, 5) (4, 4) (5, 3) (6, 2)	5
9	(3, 6) (4, 5) (5, 4) (6, 3)	4
10	(4, 6) (5, 5) (6, 4)	3
11	(5, 6) (6, 5)	2
12	(6, 6)	1

The probabilities of the 11 occurrences are $\frac{1}{36}, \frac{2}{36}, \frac{3}{36}, \frac{4}{36}, \frac{5}{36}, \frac{6}{36}, \frac{5}{36}, \frac{4}{36}, \frac{3}{36}, \frac{2}{36}$ and $\frac{1}{36}$. The entropy is therefore

$$-\frac{1}{36} \times \log_2 \frac{1}{36} - \frac{2}{36} \times \log_2 \frac{2}{36} - \frac{3}{36} \times \log_2 \frac{3}{36} - \frac{3}{36} \times \log_2 \frac{4}{36} - \frac{5}{36} \times \log_2 \frac{5}{36} - \frac{6}{36} \times \log_2 \frac{6}{36}$$

$$-\frac{5}{36} \times \log_2 \frac{5}{36} - \frac{4}{36} \times \log_2 \frac{4}{36} - \frac{3}{36} \times \log_2 \frac{3}{36} - \frac{2}{36} \times \log_2 \frac{2}{36} - \frac{1}{36} \times \log_2 \frac{1}{36} = 2.711 \text{bits.}$$

(d) *On average, how much information does the sum give about the individual scores of the dice?*

Let X be the score on the individual dice, and Y be the sum of the scores. This means that $H(X)$ is the entropy of the individual scores, and $H(X|Y)$ is the entropy that remains about the individual scores given the sum.

If the sum is 2 or 12, there is only one possible outcome, so the remaining entropy is 0. For a sum of 3, there are two equally likely possibilities, (2, 1) and (1, 2), so the remaining entropy is 1, and so on. The results for each possible sum is as follows:

Sum	No of occurrences	Remaining entropy
2	1	0
3	2	$2 \times \frac{1}{2}\log_2 \frac{1}{2} = 1$
4	3	$3 \times \frac{1}{3}\log_2 \frac{1}{3} = 1.585$
5	4	$4 \times \frac{1}{4}\log_2 \frac{1}{4} = 2$
6	5	$5 \times \frac{1}{5}\log_2 \frac{1}{5} = 2.322$
7	6	$6 \times \frac{1}{6}\log_2 \frac{1}{6} = 2.585$
8	5	$5 \times \frac{1}{5}\log_2 \frac{1}{5} = 2.322$
9	4	$4 \times \frac{1}{4}\log_2 \frac{1}{4} = 2$
10	3	$3 \times \frac{1}{3}\log_2 \frac{1}{3} = 1.585$
11	2	$2 \times \frac{1}{2}\log_2 \frac{1}{2} = 1$
12	1	0

The information given by knowledge of the sum is $I(X;Y) = H(X) - H(X|Y)$. If the sum is 2, $I(X;Y) = 5.170 - 0 = 5.170$bits, and this occurs for one possible outcome. If the sum is 7, $I(X;Y) = 5.170 - 2.585 = 2.585$bits, and this sum can occur for 6 possible outcomes. Calculating $I(X;Y)$ for each possible outcome and averaging gives us that the average information obtained is 3.274bits.

4. *In the television game show, Who Wants to be a Millionaire?, contestants have to pick a single correct answer from four choices A, B, C and D. At one point during the game, they are allowed to go '50:50', where two incorrect answers are removed, leaving the correct answer and one wrong answer from which to choose.*

(a) *Assuming a contestant has no idea what the correct answer is, and so considers each outcome equally likely, how much information is given by going '50:50'?*
The entropy before '50:50' is played is $H(X) = -\sum_{i=1}^{4} \frac{1}{4} \log \frac{1}{4} = 2$ bits. After '50:50' is played, two possibilities will remain, each of which has an equal chance (i.e. the probability of each is $\frac{1}{2}$), so $H(X) = -\sum_{i=1}^{2} \frac{1}{2} \log \frac{1}{2} = 1$ bit. The information gained is the reduction in entropy, so the information is 1 bit.

(b) *Alice is playing the game and has become stuck on a question. She is fairly certain the answer is A or B, and hopes that by playing '50:50' she will remove one of these. She estimates that it is 45% likely that the answer is A, 40% likely the answer is B, 10% likely the answer is C and 5% likely the answer is D. Calculate the entropy based on her assumptions before and after '50:50' is played if going '50:50' removes answers B and D.*
The entropy before '50:50' is played is

$$H(X) = -0.45\log(0.45) - 0.40\log(0.40) - 0.10\log(0.10) - 0.05\log(0.05) = 1.595.$$

After '50:50' is played, answers A and C remain. The probability of A is now $\frac{0.45}{0.45+0.10}$, and the probability of C is $\frac{0.10}{0.45+0.10}$, so the entropy after '50:50' is played is

$$H(X) = -\frac{0.45}{0.45+0.10}\log(\frac{0.45}{0.45+0.10}) - \frac{0.10}{0.45+0.10}\log(\frac{0.10}{0.45+0.10}) = 0.684.$$

(c) *Calculate the entropy based on her assumptions before and after '50:50' is played if going '50:50' removes answers C and D.*

The entropy before is the same as it was in the previous case at 1.595 bits. After '50:50' is played, answers A and B remain. The probability of A is now $\frac{0.45}{0.45+0.40}$, and the probability of B is $\frac{0.40}{0.45+0.40}$, so the entropy after '50:50' is played is

$$H(X) = -\frac{0.45}{0.45 + 0.40}\log(\frac{0.45}{0.45 + 0.40}) - \frac{0.40}{0.45 + 0.40}\log(\frac{0.40}{0.45 + 0.40}) = 0.9975.$$

(d) *What is the information (in bits) given by going '50:50' in each of the above two cases?*
In each case the information is the reduction in entropy, so the information given in the first case is 0.911 bits and in the second case is 0.598 bits. The first case gives more information because one of the more likely entries was removed leaving a fairly clear choice.

5. *By expanding $H(A, B)$ in terms of probabilities and using Bayes' Theorem, show that*
$H(A, B) = H(A|B) + H(B).$

$$H(A, B) = -\sum_j \sum_k P(A = a_j, B = b_k) \log_2 P(A = a_j, B = b_k)$$

Expanding the log using Bayes' Theorem gives

$$H(A, B) = -\sum_j \sum_k P(A = a_j, B = b_k) \log_2 (P(A = a_j | B = b_k) P(B = b_k))$$

$$H(A, B) = -\sum_j \sum_k P(A = a_j, B = b_k) \log_2 (P(A = a_j | B = b_k))$$
$$- \sum_j \sum_k P(A = a_j, B = b_k) \log_2 (P(B = b_k))$$

Taking the second term, $\log_2 (P(B = b_k))$ is independent of j. It can therefore be rewritten as

$$-\sum_k \left(\sum_j P(A = a_j, B = b_k) \right) \log_2 P(B = b_k)$$

$\sum_j P(A = a_j)$ includes all possibilities and therefore

$$\sum_j P(A = a_j, B = b_k) = P(B = b_k)$$

The second term is therefore

$$-\sum_k P(B = b_k) \log_2 P(B = b_k) = H(B)$$

Returning to the first term

$$-\sum_j \sum_k P(A = a_j, B = b_k) \log_2 (P(A = a_j | B = b_k))$$

Expanding using Bayes's Theorem again gives

$$
-\sum_j \sum_k (P(A = a_j | B = b_k) P(B = b_k)) \log_2 (P(A = a_j | B = b_k))
$$

$$
= -\sum_k P(B = b_k) \sum_j (P(A = a_j | B = b_k)) \log_2 (P(A = a_j | B = b_k))
$$

$$
= -\sum_k P(B = b_k) \sum_j (P(A = a_j | B = b_k)) \log_2 (P(A = a_j | B = b_k))
$$

The second part, summed over j is $H(A|B = b_k)$, the entropy of A given a particular value of B. This then gives $-\sum_k P(B = b_k) H(A|B = b_k)$, which is now summed over all the possible b_k, giving $H(A|B)$ as required.
Therefore, $H(A, B) = H(A|B) + H(B)$.

8.1.2   Answers to Questions on Memoryless Source Coding

1. *A memoryless information source produces 8 different symbols with respective probabilities of 1/2, 1/4, 1/8, 1/16, 1/32, 1/64, 1/128, 1/128. These symbols are encoded as 000, 001, 010, 011, 100, 101, 110, 111 respectively.*

   (a) *What is the entropy per source symbol?*
   The entropy per source symbol is
   $H(U) = -\sum_{k=1}^{8} p(u_i) \log p(u_i) =$
   $-\frac{1}{2} \log \frac{1}{2} - \frac{1}{4} \log \frac{1}{4} - \frac{1}{8} \log \frac{1}{8} - \frac{1}{16} \log \frac{1}{16} - \frac{1}{32} \log \frac{1}{32} - \frac{1}{64} \log \frac{1}{64} - \frac{1}{128} \log \frac{1}{128} -$
   $\frac{1}{128} \log \frac{1}{128} = 1.98$ bits.

   (b) *What is the efficiency of this code?*
   We have efficiency $= L_{min}/L$ where L is the average length of the code words.
   $L_{min} = \frac{H}{\log_2 r} = \frac{1.98}{\log_2 2} = 1.98$ $L = 3$, so efficiency $= 0.66$.

   (c) *Design a code using the Shannon-Fano algorithm, and calculate its efficiency.*
   (d) *Design a code using the Huffman algorithm, and calculate its efficiency.*
   Both methods give the same code, as follows

Symbol	Probability								L	L×Prob.
u_1	$\frac{1}{2}$	0							1	64/128
u_2	$\frac{1}{4}$	1	0						2	64/128
u_3	1/8	1	1	0					3	48/128
u_4	1/16	1	1	1	0				4	32/128
u_5	1/32	1	1	1	1	0			5	20/128
u_6	1/64	1	1	1	1	1	0		6	12/128
u_7	1/128	1	1	1	1	1	1	0	7	7/128
u_8	1/128	1	1	1	1	1	1	1	7	7/128
										254/128

   The efficiency is the minimum length over the average length $= 1.98/1.98 = 1$ (or 100%)

   (e) *If the source symbol rate is 1000/s, on average what is the encoded bit rate?*

The average symbol length is 1.98 bits for each code. The bit rate is therefore $1000 \times 1.98 = 1980$ bit/s.

2. *A memoryless information source generates symbols with probability 0.65, 0.2, and 0.15.*

   (a) *Calculate the entropy per symbol.*

   The entropy per source symbol is

   $H(U) = -\sum_{k=1}^{3} p(u_i) \, logp(u_i) = -0.65 \log 0.65 - 0.2 \log 0.2 - 0.15 \log 0.15 = 0.40397 + 0.46439 + 0.41055 = 1.2789$ bits.

   (b) *Calculate the probabilities of all possible messages consisting of two symbols, and the corresponding entropy. Compare this to the previous result.*

   Let the three symbols be u_1, u_2 and u_3. There are nine possible combinations of two symbols. These are

   $p(v_1) = p(u_1, u_1) = p(u_1)p(u_1) = 0.65 \times 0.65 = 0.4225$
   $p(v_2) = p(u_1, u_2) = p(u_1)p(u_2) = 0.65 \times 0.2 = 0.13$
   $p(v_3) = p(u_1, u_3) = p(u_1)p(u_3) = 0.65 \times 0.15 = 0.0975$
   $p(v_4) = p(u_2, u_1) = 0.2 \times 0.65 = 0.13$
   $p(v_5) = p(u_2, u_2) = 0.2 \times 0.2 = 0.04$
   $p(v_6) = p(u_2, u_3) = 0.2 \times 0.15 = 0.03$
   $p(v_7) = p(u_3, u_1) = 0.15 \times 0.65 = 0.0975$
   $p(v_8) = p(u_3, u_2) = 0.15 \times 0.2 = 0.03$
   $p(v_9) = p(u_3, u_3) = 0.15 \times 0.15 = 0.0225$

Symbol	Probability $p(v_i)$	$\log_2$ (*probability*)	$p(v_i) \log_2(p(v_i))$
v_1	0.4225	−1.2430	−0.5252
v_2	0.13	−2.9434	−0.3826
v_3	0.0975	−3.3585	−0.3275
v_4	0.13	−2.9434	−0.3826
v_5	0.04	−4.6439	−0.1858
v_6	0.03	−5.0589	−0.1518
v_7	0.0975	−3.3585	−0.3275
v_8	0.03	−5.0589	−0.1518
v_9	0.0225	−5.4739	−0.1232
Sum of $p(v_i) \log_2(p(v_i))$			−2.558

   The entropy is therefore 2.558 bits.

   Regarding the comparison, we can either calculate the entropy as the sum over all the nine events, or we can note that as the symbols are independent, the entropy of two symbols is twice that of one symbol (and in general, a message with l symbols has entropy $lH(U)$).

   $H(U)$ is 1.2789 bits (see above). l is 2, which gives 2.5578 bits, which is the same. Note that the result is accurate to 4 significant figures since the internal calculations used 5 significant figures.

   (c) *Calculate the redundancy of the information source.*

   The redundancy is equal to $\left(1 - \dfrac{H(U)}{\max H(U)}\right)$

   The maximum value of $H(U)$ is when each symbol is equally likely (i.e. has probability 1/3). This has value $\log_2 3$. Redundancy is therefore $1 - (1.2789/\log_2 3) = 0.1931$, or 19.3%.

3. *Design a code to encode a memoryless source with six symbols using a tertiary encoded alphabet $\{0,1,2\}$. A sample of the source output is ACAAABEBCDAEABDCAEFAABDF. What is the efficiency of your code? Compare it to the code $\{A:00, B:01, C:02, D:10, E:11, F:12\}$*

From the string ACAAABEBCDAEABDCAEFAABDF, A occurs 9 times, B occurs 4 times, C occurs 3 times, D occurs 3 times, E occurs 3 times, and F occurs twice. Assuming that this string is representative, the probabilities of the symbols are A= $\frac{9}{24} = \frac{3}{8}$, B= $\frac{4}{24} = \frac{1}{6}$, C= $\frac{3}{24} = \frac{1}{8}$, D= $\frac{3}{24} = \frac{1}{8}$, E= $\frac{3}{24} = \frac{1}{8}$, and F= $\frac{2}{24} = \frac{1}{12}$, The Shannon-Fano method for a ternary code proceeds by constantly dividing the symbols into three groups of approximately equal probabilities. In this manner the following code arises:

Symbol	Probability				L	LProb
A	3/8=9/24	0			1	9/24
B	1/6=4/24	1	0		2	8/24
C	1/8=3/24	1	1		2	6/24
D	1/8=3/24	2	0		2	6/24
E	1/8=3/24	2	1		2	6/24
F	1/12=2/24	2	2		2	4/24
						1.625

$$H(U) = -\sum_{k=1}^{6} p(s_i) \log p(s_i)$$

$$= \frac{9}{24} \log \frac{9}{24} - \frac{6}{24} \log \frac{6}{24} - 3 \times \left(\frac{3}{24} \log \frac{3}{24} \right) - \frac{2}{24} \log \frac{2}{24} = 2.39 \text{ bits.}$$

$L_{min} = \frac{H}{\log_2 r} = \frac{2.39}{\log_2 3} = 1.511$ The average length, L, is the sum of the individual codeword probabilities multiplied by their length.

$$L = 1 \times \frac{9}{24} + 2 \times \frac{8}{24} + 2 \times \frac{6}{24} + 2 \times \frac{6}{24} + 2 \times \frac{6}{24} + 2 \times \frac{4}{24} = 1.625 \text{ bits.}$$

Efficiency $= L_{min}/L = 0.930$, or 93.0%.

The second code given, which which has a constant length of two symbols per message would have $L = 2$, and so an efficiency of 75.5%.

4. *Remote sensors can have two states – 'SET' and 'CLEAR'. On average, 99% of sensors are in the 'CLEAR' state.*

(a) *A one bit code, $0 \to$ CLEAR, $1 \to$ SET, is used. What is its efficiency?*

Entropy $H = -0.01 \times \log_2 0.01 - 0.99 \times \log_2 0.99 = 0.080793$ bits.

Average length of a codeword is 1 (since both codewords have length 1). The efficiency of the code is therefore $L_{min}/L_{av} = H/L_{av}$ for a binary code $= 0.080793$, or 8.08%

(b) *An improvement of the efficiency is sought by constantly taking two messages together. Determine a suitable binary code for this, and calculate its efficiency.*

Since each set/clear event is independent, entropy is the sum of the entropy of each event, so

$H = 0.080793 + 0.080793 = 0.161583$ bits.

A suitable code by the method of Shannon-Fano is:

Event	Probability				L	L×Prob.
CC	$(0.99)^2$	0.9801	0		1	0.9801
CS	$(0.99)^1(0.01)^1$	0.0099	$\overline{1}$ 0		2	0.0198
SC	$(0.99)^1(0.01)^1$	0.0099	1 $\overline{1}$ 0		3	0.0297
CC	$(0.01)^2$	0.0001	1 1 $\overline{1}$		3	0.0003
			Average length (sum of L×Prob)			1.0299

Efficiency $= H/L_{av} = 0.161583/1.0299 = 0.156894$, or 15.69%

(c) *Repeat for groups of 3 symbols.*
   H is now 3×(entropy of single event) = 0.242379.
   A suitable code by the method of Shannon-Fano is:

Event	Probability			L	L×Prob.
CCC	$(0.99)^3$	0.970299	0	1	0.970299
CCS	$(0.99)^2(0.01)^1$	0.009801	$\overline{1}$ 0 0	3	0.029403
CSC	$(0.99)^2(0.01)^1$	0.009801	1 0 $\overline{1}$	3	0.029403
SCC	$(0.99)^2(0.01)^1$	0.009801	1 $\overline{1}$ 0	3	0.029403
CSS	$(0.99)^1(0.01)^2$	0.000099	1 1 $\overline{1}$ 0 0	5	0.000495
CSC	$(0.99)^1(0.01)^2$	0.000099	1 1 1 0 $\overline{1}$	5	0.000495
CSC	$(0.99)^1(0.01)^2$	0.000099	1 1 1 $\overline{1}$ 0	5	0.000495
SCC	$(0.01)^3$	0.000001	1 1 1 1 $\overline{1}$	5	0.000005
			Average length (sum of L×Prob)		1.059998

Efficiency $= H/L_{av} = 0.242379/1.059998 = 0.2286598$, or 22.87%.

(d) *Repeat for groups of 4 symbols.*
   H is now 4×(entropy of single event) = 0.32317254.
   A suitable code by the method of Shannon-Fano is:

Event	Probability			L	L×Prob.
CCCC	$(0.99)^4$	0.96059601	0	1	0.960596
CCCS	$(0.99)^3(0.01)^1$	0.00970299	$\overline{1}$ 0 0	3	0.029109
CCSC	$(0.99)^3(0.01)^1$	0.00970299	1 0 $\overline{1}$	3	0.029109
CSCC	$(0.99)^3(0.01)^1$	0.00970299	1 $\overline{1}$ 0	3	0.029109
SCCC	$(0.99)^3(0.01)^1$	0.00970299	1 1 $\overline{1}$ 0	4	0.038881
CCSS	$(0.99)^2(0.02)^2$	0.00009801	1 1 1 $\overline{1}$ 0 0	6	0.000588
CSCS	$(0.99)^2(0.02)^2$	0.00009801	1 1 1 1 0 $\overline{1}$ 0	7	0.000686
CSSC	$(0.99)^2(0.02)^2$	0.00009801	1 1 1 1 0 1 $\overline{1}$	7	0.000686
SCCS	$(0.99)^2(0.02)^2$	0.00009801	1 1 1 1 $\overline{1}$ 0 0	7	0.000686
SCSC	$(0.99)^2(0.02)^2$	0.00009801	1 1 1 1 1 0 $\overline{1}$	7	0.000686
SSCC	$(0.99)^2(0.02)^2$	0.00009801	1 1 1 1 1 $\overline{1}$ 0	7	0.000686
CSSS	$(0.99)^1(0.01)^3$	0.00000099	1 1 1 1 1 1 $\overline{1}$ 0 0	9	0.000009
SCSS	$(0.99)^1(0.01)^3$	0.00000099	1 1 1 1 1 1 1 0 $\overline{1}$	9	0.000009
SSCS	$(0.99)^1(0.01)^3$	0.00000099	1 1 1 1 1 1 1 $\overline{1}$ 0	9	0.000009
SSSC	$(0.99)^1(0.01)^3$	0.00000099	1 1 1 1 1 1 1 1 $\overline{1}$ 0	10	0.000010
SSSS	$(0.01)^3$	0.00000001	1 1 1 1 1 1 1 1 1 $\overline{1}$	10	0.0000001
			Average length (sum of L×Prob)		1.0908591

Efficiency $= H/L_{av} = 0.32317254/1.0908591 = 0.296255$, or 29.6%.

(e) *Comment on any trends you see in the above results.*

Efficiency increases with the number of events encoded; taking several events together allows unlikely events to have longer codewords in line with their lower probabilities; combining events also results in a more complex code.

8.1.3   Answers to Questions on Source Coding for Sources with Memory

1. *Encode the following sequence using delta modulation: 1, 2, 4, 4, 4, 4, 5, 6, 6, 5, 5, 4, 3, 1, 0. Assume the decoder is outputting 0 before reception of the first bit. What is the mean squared error?*

   The first input is 1, which is above the current output (0), so the encoder sends a 1, and the output is 1. The second input is 2, which is above 1, so a 1 is again sent, and so on. The sequence is

Current input	1	2	4	4	4	4	5	6	6	5	5	4	3	1	0
Current output	0	1	2	3	4	5	4	5	6	7	6	5	4	3	2
Bit sent	1	1	1	1	1	0	1	1	1	0	0	0	0	0	0
New output	1	2	3	4	5	4	5	6	7	6	5	4	3	2	1
Error	0	0	1	0	1	0	0	0	-1	-2	0	0	0	-1	-1

   The transmitted sequence is 111110111000000. A slightly different sequence would have resulted if a 0 was transmitted if the values were equal rather than a 1. This is an implementation option; either assumption is correct.

   The mean square error is the mean value of the square of the error, i.e. the mean of (0, 0, 1, 0, 1, 0, 0, 0, 1, 4, 0, 0, 0, 1, 1) = 0.6.

2. *A source generates the sequence 15, 14, 12, 9, 8, 8, 10, 11, 12, 14, 15, 15, 14, 12, 10, 9.*

   (a) *Encode the sequence using differential PCM*

   The first value is sent as a reference, and then the differences are sent, giving 15, –1, –2, –3, –1, 0, 2, 1, 1, 2, 1, 0, –1, –2, –2, –1.

   (b) *If you are told that the values which will be received are integers in the range 0 and 15, how many bits per sample will be required to encode each value in PCM?*

   There are 16 values, so $\log_2 16 = 4$ bits are required.

   (c) *How many bits per symbol will be required in adaptive PCM? Justify your answer.*

   3 bits. From the sample, it looks as if the maximum difference between values is 3. This can be positive or negative, and a difference of 0 is also possible. Therefore, there are 7 possibilities, and 3 bits are needed per symbol (after the first one).

   Alternatively, 5 bits per symbol could be used to give a lossless scheme, since it allows all possibilities from –15 to 15. However, without further coding the result would be an expansion in the data.

   (d) *Is the system you propose lossy or lossless? If lossy, how could you make it lossless?*

   For the first proposal, lossy. Lossless would require the ability to encode a 0 after a 15 or vice versa, requiring 5 bits per symbol.

3. *The same sequence as given above is to be transmitted using adaptive PCM. The estimate used of the next sample is to assume that the difference between the current value and the next is the same as between the current and the last.*

   (a) *What are the values to be sent?*

Value	15	14	12	9	8	8	10	11	12	14	15	15	14	12	10	9
Estimate	0	15	13	10	6	7	8	12	12	13	16	16	15	13	10	8
Difference (i.e. value sent)	15	−1	−1	−1	2	1	2	−1	0	1	−1	−1	−1	−1	0	1

(b) *Estimate the number of bits required per symbol.*

In this example, only −1, 0,1 and 2 need to be sent (after the initial value). Two bits per sample could therefore suffice, although this risks −2 being required which the system could not then handle. The code designer may decide that increased error in such cases is worthwhile. Alternatively, Huffman coding could be used to compress the values. Note that adaptive differential PCM results in fewer, but more frequently occurring values. These would compress more easily with Huffman coding than the differential PCM example.

4. *Use the 1977 Lempel-Ziv algorithm to compress the sequence ABCCDABDCCBC-CBECDA. Compare the number of bits before and after compression. The sequence is 18 symbols long, so 5 bits are needed to address the string. Comment on what happens regarding addressing on long sequences. If the algorithm could only address previous strings over a finite history using a sliding window, what advantage and disadvantage would this have?*

ABCCD is sent without any coding, as there are no repetitions (C is a only one symbol long, so encoding this offers no advantages). AB is a repeat of the string starting at position 1, lasting two symbols (1,2). D is sent, as it does not form part of a repeating string. CC is a repeat of (3,2) BCC is sent as (2,3). BE is sent normally, then CDA is sent as (4,3).

There are five symbols, A, B, C, D and E. If each symbol is sent using a binary encoding, 3 bits are required (assuming no compression is used). A repeating string requires 5 bits for the address, and 2 bits for the length, in this example. This is 7 bits. Therefore, in this example, it would not be worthwhile encoding repeating strings of two symbols since the encoded form is one bit longer than the plain text form. Some compression is given when three symbol strings are repeated (2 bits).

If a sliding window was used, this would reduce the addressing requirements, and limit the number of bits needed in encoded form. However, the system would not then be able to retrace over the full length of the transmission, and may miss opportunities for encoding repetitions.

8.2   Answers to Questions on the Security Perspective

1. *Using a transposition cipher with a grid depth of 8, encode the message 'Takeover announcement will be made at noon tomorrow'.*

The grid is as follows:

T	A	K	E	O	V	E	R
A	N	N	O	U	N	C	E
M	E	N	T	W	I	L	L
B	E	M	A	D	E	A	T
N	O	O	N	T	O	M	O
R	R	O	W				

The message is therefore TAMBNRANEEORKNNMOOEOTANWOUWDT VNIEO ECLAM RELTO.

2. *Repeat the encoding using a transposition cipher with the keyword 'COLUMNAR' to define the ordering of the columns.*
   The alphabetical ordering of the letters in 'columnar' is 26384517, so we read the columns in that order, giving ECLAM TAMBNRKNNMOOOUWDT VNIEO ANEEORRELTO EOTANW.

3. *Encode 'Now is the time for all good men to come to the aid of the party' using Caesar cipher with a shift of 4.*
   RSA MW XLI XMQI JSV EPP KSSH QIR XS GSQI XS XLI EMH SJ XLI TEVXC

4. *Decrypt KWWS://ZZZ.DQWHQQD-PRGHOV.FRP/VHFUHW/LQGH0.KWPO.*
   This looks like a web address. If it is, KWWS is 'http', and ZZZ is likely to be 'www'. In fact, this would imply K is h, W is t, and Z is w, all of which conform to a Caesar substitution with a shift of 3, so the solution follows directly.

   However, assuming that an arbitrary letter ordering has been used, with the above assumptions we have:

   http://www.DQtHQQD-PRGHOV.FRP/VHFUHt/LQGHA.htPO

   This would suggest P is m and O is l. The first part is a domain name, with a three letter top level, the last letter of which is likely to be m. This is therefore probably 'com'. We now have:

   http://www.DQtHQQD-moGHlV.com/VHcUHt/LQGHA.html

   We don't have enough letters to guess the domain name, so we have to make another guess. The initial pages of a web site are often 'main' or 'welcome' but the most common is 'index'. This fits in terms of numbers of letters, and gives us:

   http://www.DntennD-modelV.com/VecUet/index.html

   V looks like being s, giving secUet, which is probably 'secret' (which is a very bad word for a password!). This leaves DntennD. We could try all remaining letters to find this, or notice that the only letter which makes that a word is 'a'. Domain names don't have to be words, but most are, since they are easier to remember. A little typing at a web browser confirms that this is the correct decryption.

5. *Using the Vigenère cipher and the key word 'hide', encode 'We must meet under the clock.'*
   DM PYZB PILB XRKMU XOM FPVKN

6. *Using the stream cipher and an initial key of 'D', encode 'The documents will be sent tomorrow.'*
   WDH KYAUGKXQI EMXI JN FJWP IWIWNESO

7. *What is the bit stream generated by the following LFSR?*

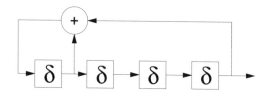

Starting with the shift registers in state 1000 gives the following sequence: 1000, 1100, 1110, 1111, 0111, 1011, 0101, 1010, 1101, 0110, 0011, 1001, 0100, 0010, 0001, and back to 1000. Therefore the output is 0001111010110010.

8. *For an RSA system with primes $p = 7$ and $q = 13$, with encryption key $e = 5$, find the following.*

   (a) *What is the value of the private key d?*
      $ed = 1$ (mod $(p - 1)(q - 1)$). By trying successive numbers, $d = 29$, since $5 \times 29 = 145 \bmod 72 = 1$.
   (b) *Encrypt the number 2 using your system, and verify that it is possible to decrypt the resulting codeword back to the message.*
      The message 2 is encoded by raising it to the power of 5 and then taking the result modulo 91. This gives us that $2^5 \bmod 91 = 32$
      To decrypt 32, it is necessary to take it to the power of 29, and calculate this modulo 91. Most calculators cannot do this to the necessary precision, so we proceed as follows:
      $32^{29} \bmod 91 = 32^5 32^5 32^5 32^5 32^5 32^4 \bmod 91 = 2 \times 2 \times 2 \times 2 \times 74 \bmod 91 = 2$

8.3   Answers to Questions on the Network Perspective

8.3.1   Answers to Questions on Network Configuration

1. *A communications network consists of N nodes. For the following cases, determine the number of links required to connect these nodes assuming that each link is bi-directional.*

   (a) *A 5 node ring*
      N node ring requires N bi-directional links – 5 nodes requires 5 links.
   (b) *An 8 node fully interconnected mesh*
      N node mesh requires $N(N - 1)/2$ bi-directional links – 8 nodes requires 28 links.
   (c) *A 5-connected, 10 node partially interconnected mesh*
      5 connected network means that each node has a degree of 5 and is connected to 5 other nodes. Total number of links required $= 5 \times 10 \times \frac{1}{2} = 25$. Contrast with a fully interconnected 10 node network that requires 45 links.
   (d) *Repeat the above three cases when $N = 96$.*
      $N = 97$, ring requires 96 links, mesh requires 4560 nodes and a 5 connected 96 node network needs 240 links.

2. *Determine the connectivity of the two networks given in the following two diagrams. For both networks, how many iterations are required to determine the connectivity?*

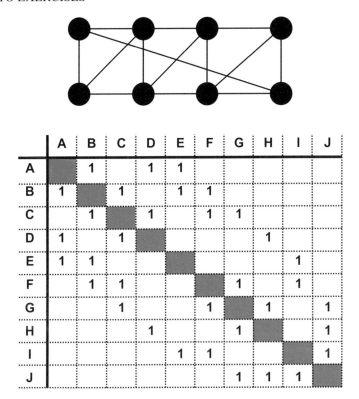

	A	B	C	D	E	F	G	H	I	J
A		1		1	1					
B	1		1		1	1				
C		1		1		1	1			
D	1		1					1		
E	1	1							1	
F		1	1				1		1	
G		1				1		1		1
H				1			1			1
I					1	1				1
J							1	1	1	

In both networks, the minimum degree is 3, thus the connectivity cannot exceed 3. One network has 8 nodes while the other has 10 nodes but both have connectivity of 3.

N nodes in the network and the minimum degree is M.

Evens:
1st part requires $M!/2$ path searches followed by $N - M$ searches. This gives a total of $(\frac{1}{2}M!) + N - M)$ searches. This is 8 and 10 for each network.

Kleitmans:
Require $N - 1 + N - 2 + \ldots + N - M$ searches. This is 18 and 24 respectively for each network.

8.3.2 Answers to Questions on Switching Techniques

1. *A 64 kbit/s circuit-switched path between two users utilising 9 switching stages takes 5 seconds to be set up. How long will it take, in total, to directly transfer the entire contents of a 1.4 Mbyte floppy disk between the two users using this channel?*
   Time to transfer data = amount of data / bit-rate = 1.4Mbytes /64000 = 183.5 seconds.
   Total delay = transfer time + set-up time = 183.5 + 5 = 188.5 seconds

2. *If the above circuit-switched path is replaced by a 9 node-packet switched system utilising 1 Mbit/s links, how long does it take to transfer the 1.4 Mbyte file, assuming that there is no delay at any of the switching nodes and that packet payload is 256 octets and packet overheads are equivalent to 18 octets?*

1 packet takes $(256 + 18) \times 8/10^6 = 2.192$ ms to pass over 1 stage. There are $9 + 1$ stages to pass through, which gives a 21.92 ms end-to-end delay for 1 packet.

A 1.4 Mbyte file is equivalent to 5735 packets to be carried. Total delay to transfer file is then the total delay to carry 1 packet over the 10 stages plus the time to carry the remaining 2734 packets over the last stage. Total delay = 12.591 seconds.

3. *Contrast the delays when packet payload is changed to 128, 512, 1024 and 4096 octets respectively.*

   When packet size is reduced to 128 octets, link delay becomes 1.168 ms per packet but we now require to send 11,469 packets. Total delay becomes 13.4 seconds.

   When packet size is increased to 512 octets, link delay becomes 4.24 ms per packet but we now require to send 2,868 packets. Total delay becomes 12.198 seconds.

   When packet size is increased to 1024 octets, link delay becomes 8.336 ms per packet but we now require to send 1434 packets. Total delay becomes 12.028 seconds.

   When packet size is increased to 4096 octets, link delay becomes 32.912 ms per packet but we now require to send 359 packets. Total delay becomes 12.111 seconds.

8.3.3   Answers to Questions on Network Dimensioning

1. *A simple telephone exchange consists of 4 circuits. If, at the busiest time of day, calls arrive at the exchange every 240 seconds, calculate the occupancy of the exchange and determine the probability that a call will be rejected.*

   In this case $A = 120/240 = 0.5$. There are 4 circuits. Solution can be found directly by plugging in numbers into Erlang B equation or solving iteratively. $E(A, n) = AE(A, n - 1)/(AE(A, n - 1) + n)$.

   $E(0.5, 0) = 1, E(0.5, 1) = 0.5/(0.5+1) = 1/3, E(0.5, 2) = 0.5/3/(0.5/3+2) = 0.0769$
   Finally $E(0.5, 4) = 1.579 \times 10^{-3}$. This means that 99.84% of all calls are carried and exchange occupancy is $0.9984 \times 0.5 = 0.4992$ per circuit.

2. *If, in the above system, the users are happy with a loss of 2%, determine the number of circuits required in the multiplexor.*

   $E(0.5, 3) = 0.0127 = 1.27\%$ loss. Thus it would be possible to have 3 circuits in this system and still provide a loss below 2%.

3. *If the call rate in the above system increases fivefold, determine the call loss probability and utilisation of each of the 4 exchange circuits.*

   If call rate increased fivefold, new traffic would be 2 Erlangs. The $E(2, 4)$ would give 15% loss. Carried traffic would be $2.5 \times 0.85 = 2.125$ as 15% is dropped. Circuit utilisation is 0.53125, $\frac{1}{4}$ of total carried traffic.

4. *A terminal concentrator consists of six 56 kbit/s input lines and a single 128 kbit/s output line. The mean packet size is 450 bytes and the arrival rate associated with each input line is 5 packets/sec. What is the mean delay experienced by a packet and what is the mean number of packets stored in the concentrator?*

   $\mu = \frac{128000}{450/8} = 35.56$ packets per second

   $\lambda = 6 \times 5 = 30$ packets per second

   $\rho = \lambda/\mu = 30/35.6 = 0.844$

   Mean Delay $T = \frac{1}{\mu - \lambda} = \frac{1}{35.56 - 30} = 179.8$ ms.

   Occupancy $= \frac{\rho}{1-\rho} = \frac{0.844}{1 - 0.844} = 5.4$

5. *Two computers are inter-connected via a 64-kbit/s line and currently support 8 interactive sessions (connections). If the mean packet length is 150 bits and the arrival rate/session is 4 packet/s, should the network provide each session with its own dedicated 8 kbit/s channel or should all sessions compete for the entire line capacity when packet delay is the most important criterion?*

Shared Capacity

$\mu = \frac{64000}{150/8} = 53.3$ packets per second

$\lambda = 8 \times 4 = 32$ packets per second

$\rho = \lambda/\mu = 32/53.3 = 0.6$

Mean Delay $T = \frac{1}{\mu - \lambda} = \frac{1}{53.33 - 32} = 46.88$ ms.

Single Channels – 1 channel of 8 kbit/s supporting 4 packets per second

$\mu = \frac{8000}{150/8} = 6.67$ packets per second

$\lambda = 4$ packets per second

$\rho = \lambda/\mu = 4/6.67 = 0.6$

Mean Delay $T = \frac{1}{\mu - \lambda} = \frac{1}{6.67 - 4} = 375$ ms.

It is better to share the capacity as delay is 1/8 of dedicated channel. Utilisation is the same in both cases.

8.3.4 Answers to Questions on Routing

1. *Routing within a 6 node network is based upon a distance vector approach; router A is adjacent to routers B, D and E. The distance vectors associated with each of these 3 routers are given below. Using these tables, derive a new routing table for node A. Assume that the current distance between node A and its neighbours is accurately reflected in the tables.*

To answer this question it is necessary to determine routes from A to C and A to F as all other nodes are adjacent to node A. The right-hand table represents the answer to the question.

NODE B		NODE D		NODE E		NODE A	
A	10	A	20	A	15	A	-
B	-	B	15	B	10	B	B(10)
C	20	C	40	C	25	C	B(30)
D	30	D	-	D	10	D	D(20)
E	10	E	10	E	-	E	E(15)
F	10	F	25	F	5	F	E(20)

2. *The current link state tables for a 6 node network are given below. Draw this network and determine the routing table associated with node A.*

Use the given table (repeated below) and draw the network.

A	B(10)	D(5)	E(20)	
B	A(10)	C(20)	E(10)	F(10)
C	B(20)	F(20)		
D	A(5)	E(10)		
E	A(20)	B(10)	D(10)	F(10)
F	B(10)	C(20)	E(10)	

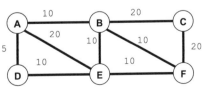

Obtain shortest path from A to all other nodes by inspection or use formal algorithm.

NODE A	
A	-
B	B(10)
C	B(30)
D	D(5)
E	D(15)
F	B(30)

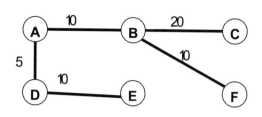

3. *Router A is required to broadcast a packet to all other routers within this network. If the packet field contains a hop count initialised to the diameter of the network, determine the minimum and maximum number of packets this network may be required to transmit in order to convey this broadcast packet.*

The minimum number is 5 (the number of links in spanning tree). The maximum number is 24 (flooded packets with hop count of 3).

8.3.5   Answers to Questions on Error Control

1. *A data link protocol has the following characteristics:*

Data length	100 bytes
Header length	8 bytes
Channel capacity	2 Mbit/s
Acknowledgement frame length	8 bytes
Probability of lost frame	2×10^{-3}
Probability of lost acknowledgement	5×10^{-4}
Service and propagation delay	0.15 ms

(a) *Estimate the maximum possible efficiency of this system if the protocol operates in a simple 'stop and wait' manner using positive acknowledgements. What is the impact of setting a timeout interval of 1 ms?*

$$\text{Efficiency,} = \frac{m}{m+h+a+2B\tau} = \frac{100}{(100+8+8)8+2 \times 2000 \times 0.15} = 0.524$$

Timeout interval $T = (m+h+a)/B+2\tau = (100+8+8) \times 8/2000+2 \times 0.15 = 0.764$, setting to 1 ms would mean that efficiency is lowered as each lost frame would incur a delay of 1 ms rather than the optimum of 0.764 ms. $P_s = (1 - P_f)(1 - P_a) = (1 - 2 \times 10^{-3})(1 - 5 \times 10^{-4}) = 0.9975$. This means that around 0.0025 of all packets require retransmissions – this is not a significant number and does not significantly affect the overall efficiency of the link.

(b) *In order to maximise link efficiency, the 'stop and wait' protocol is to be changed into a sliding window protocol. Estimate an appropriate window size, if in the first instance the link is assumed to be error free.*

$Wm/B > (m+h+a)/B + 2\tau$ gives the threshold for window size, solve to get W. $W > (m+h+a+2B\tau)/m = (100+8+8) \times 8+2 \times 2000 \times 0.15)/800 = 1.91$. This means W must be at least 2.

2. *The key parameters of a data link and its ARQ protocol are given in the table below. If a 3-bit SEQ field is employed, which retransmission mechanism gives optimal link performance?*

Information field	150 bytes
Frame overhead	16 bytes
Acknowledgement length	16 bytes
Probability of lost frame	10^{-3}
Probability of lost acknowledgement	2×10^{-4}
SEQ field size	3 bits
Acknowledgement time out	150 ms
Link bit rate	64 kbit/s
Link propagation delay	50 ms

A window size of W requires a sequence number range of W for go back N compared to $2W$ for selective. The given SEQ field means that sequence numbers can range from 0–7. For go back N, the effective window size is 8.

For selective, the effective window size is 4.

We need to determine for both schemes whether windows are large or small.

Time to transmit and ack 1 frame= $\frac{m+h}{B} + \frac{a}{B} + 2\tau = \frac{(150+16+20)8}{64} + 2 \times 50 = 123.25$ ms

Time to Transmit $W = 4$ frames= $\frac{W(m+h)}{B} = \frac{4(150+16)8}{64} = 83$ ms, so it is a small window.

Time to Transmit $W = 8$ frames= $\frac{W(m+h)}{B} = \frac{8(150+16)8}{64} = 166$ ms, so it is a large window.

Consider the effects of errors:

$P_s = (1 - P_f)(1 - P_a)$ where P_f is the probability that a frame is lost and P_a is the probability that an ACK is lost. In this case:

$P_s = 1 - (1 - 10^{-3})(1 - 2 \times 10^{-4}) = 0.9988002$

This is sufficiently close to unity to ignore. Also common to both efficiency equations.

Go back N

Window Size $(W) = 8$. Continuous

Efficiency is found from the following equation

Efficiency= $\frac{m}{m+h} P_s = 150166 \times 0.9988002) = 0.903$

Selective

Window Size $(W) = 4$. Not continuous.

Efficiency= $\frac{mWP_s}{m+h+a+2B\tau} = \frac{150 \times 8 \times 4 \times 0.9988}{(150+16+16) \times 8} + 2 \times 64 \times 50 = 0.6795$.

3. *A dual speed point-to-point communication link utilises a dual mode data link protocol that is capable of being switched between a forward error control mechanism or an ARQ protocol based upon a go back N retransmission scheme. This link operates in one of two speeds – low and high – and has key characteristics as shown in the table below. The link's FEC mechanism is characterised by a 50% redundancy and perfect error correction. Which error control mode should be applied to each link speed if the key aim is to maximise link efficiency?*

Information field	200 bytes
Frame overhead	16 bytes
Acknowledgement length	20 bytes
Probability of lost frame	10^{-3}
Probability of lost acknowledgement	2×10^{-4}
Window size	8
Acknowledgement time out	250 ms
Low link bit rate	32 kbit/s
High link bit rate	256 kbit/s
Link propagation delay	100 ms

FEC: 100% overhead means 1 bit of information requires the transmission of 2 bits of data – redundancy of 50%. Thus efficiency of data link is 50% regardless of the link bit-rate.

BEC: ARQ – go back N

Consider the effects of errors – this can be done at a variety of points in the solution with a variety of interpretations.

$P_s = (1 - 10^{-3})(1 - 2 \times 10^{-4}) = 0.9988002$

This is sufficiently close to unity to ignore. Including this value will have little effect numerically on the final conclusion.

BEC – High Speed

Window Size (W) = 8. We need to determine if this is sufficient for continuous transmission or not.

Time to Transmit W frames= $\frac{W(m+h)}{B} = \frac{8 \times (200+16) \times 8}{256} = 54$ ms

Time to transmit and ack 1 frame= $\frac{m+h}{B} + \frac{a}{B} + 2\tau = \frac{(200+16+20) \times 8}{256} + 2 \times 100 = 207.37$ ms

Thus in this case, transmission is not continuous and transmitter will need to stop and wait for ACKs to return before continuing to transmit frames. Therefore the window is small.

Eff= $\frac{mWP_s}{m+h+a+2B\tau} = \frac{200 \times 8 \times 8 \times 0.9988002}{(200+16+20) \times 8} + 2 \times 256 \times 100 = 0.24$

BEC – Low Speed

Window Size (W) = 8. We need to determine if this is sufficient for continuous transmission or not.

Time to Transmit W frames= $\frac{W(m+h)}{B} = \frac{8 \times (200+16) \times 8}{32} = 432$ ms

Time to transmit and ack 1 frame= $\frac{m+h}{B} + \frac{a}{B} + 2\tau = \frac{(200+16+20)8}{3}2 + 2 \times 100 = 259$ ms

Thus in this case, transmission is continuous – transmitter will not need to stop and wait for ACKs to return before continuing to transmit frames. Therefore the window is large.

Efficiency is equal to $\frac{m.P_s}{m+h} = 200 \times 0.9988002216 = 0.926$

Summary

The following figures apply:

	FEC	BEC
Low Speed	50%	92.6%
High Speed	50%	24%

Thus FEC should be employed with high speed link and BEC with low speed link.

8.4   Answers to Questions on the Link Perspective

8.4.1   Answers to Questions on Linear Codes

1. *What is the probability that a 4 bit message will be received correctly in a system using a (7, 4) Hamming code if the probability of a bit error is 2%?*
   A Hamming code can correct one error. The message will be received correctly if 0 errors or one error occurs. This is given by:
   $P(0 \text{ error}) = (0.98)^7 = 0.8681$
   $P(1 \text{ error}) = 7(0.98)^6(0.02)^1 = 0.1240$
   Therefore, the probability of correct reception is $0.8681 + 0.1240 = 0.992$ (or 99.2%)

2. *Which is more likely: More than one error when transmitting 16 bits through a BSC (Binary Symmetric Channel) with transition probability 0.1, or more than two errors when transmitting 8 bits through a channel with transition probability 0.2?*
   Case 1 (more than one error is 16 bits) :
   $$\begin{aligned} P(\text{more than 1 error}) &= 1 - P(0) - P(1) \\ &= 1 - (0.9)^{16} - 16(0.9)^{15}(0.1)^1 \\ &= 1 - (0.1853) - (0.3294) \\ &= 0.4853 \end{aligned}$$
   Case 2 (more than two errors in 8 bits) :
   $$\begin{aligned} P(\text{more than 2 errors}) &= 1 - P(0) - P(1) - P(2) \\ &= 1 - (0.8)^8 - 8(0.8)^7(0.2)^1 - \tbinom{8}{2}(0.8)^6(0.2)^2 \\ &= 1 - (0.8)^8 - 8(0.8)^7(0.2)^1 - 28(0.8)^6(0.2)^2 \\ &= 1 - (0.1678 + 0.3355 + 0.2936) \\ &= 1 - 0.7969 = 0.203 \end{aligned}$$
   Therefore, case 1 is more likely.

3. *What is the minimum possible number of parity bits theoretically required for a two error correcting code which is 16 bits long?*
   This can be found by working out the number of possible combinations of errors.
      One 'combination' is that no errors occur The number of possible combinations of one error is 16 (since the error can occur in one of 16 positions).
      The number of possible combinations of two errors is $\binom{16}{2} = \frac{16!}{14!2!} = \frac{16 \times 15}{2 \times 1} = 120$
      There are therefore 137 combinations of up to 2 errors. If the error correcting code is to work, there must be a distinct parity bit pattern for each event. Another way of saying this is to say that the parity bits must be able to 'encode' the information from the error combinations. Using Hartley's Theorem, the information content of 137 events is $\log_2 137$ bits, or a little over 7. Therefore, at least 8 binary digits are required for the parity. The minimum number of parity bits is 8.

4. *Put the following matrix into standard echelon form*

$$\begin{pmatrix} 0 & 0 & 0 & 1 & 0 & 1 & 1 \\ 0 & 0 & 1 & 0 & 1 & 1 & 0 \\ 0 & 1 & 0 & 1 & 1 & 0 & 0 \\ 1 & 0 & 1 & 1 & 0 & 0 & 0 \end{pmatrix}$$

Stage one: Reverse the ordering of the rows. New rows 3 and 4 are now okay.

$$\begin{pmatrix} 1 & 0 & 1 & 1 & 0 & 0 & 0 \\ 0 & 1 & 0 & 1 & 1 & 0 & 0 \\ 0 & 0 & 1 & 0 & 1 & 1 & 0 \\ 0 & 0 & 0 & 1 & 0 & 1 & 1 \end{pmatrix}$$

Stage two: Add rows 3 and 4 to row 1 to remove the ones in columns 3 and 4.

$$\begin{pmatrix} 1 & 0 & 0 & 0 & 1 & 0 & 1 \\ 0 & 1 & 0 & 1 & 1 & 0 & 0 \\ 0 & 0 & 1 & 0 & 1 & 1 & 0 \\ 0 & 0 & 0 & 1 & 0 & 1 & 1 \end{pmatrix}$$

Stage three: Add row 4 to row 2 to remove the 1 in column 4.

$$\begin{pmatrix} 1 & 0 & 0 & 0 & 1 & 0 & 1 \\ 0 & 1 & 0 & 0 & 1 & 1 & 1 \\ 0 & 0 & 1 & 0 & 1 & 1 & 0 \\ 0 & 0 & 0 & 1 & 0 & 1 & 1 \end{pmatrix}$$

5. *Form the generator matrix of a (15,11) Hamming code and its corresponding parity check matrix.*

The generator matrix is of a Hamming code is formed from a $k \times k$ identity matrix, followed by all combinations of $n - k$ bit vectors with 2 or more 1s. Different orderings will result in different, but equivalent codes. One possible generator matrix is

$$G = \begin{pmatrix} 1 & 0 & 0 & 0 & 0 & 0 & 0 & 0 & 0 & 0 & 0 & 1 & 1 & 0 & 0 \\ 0 & 1 & 0 & 0 & 0 & 0 & 0 & 0 & 0 & 0 & 0 & 1 & 0 & 1 & 0 \\ 0 & 0 & 1 & 0 & 0 & 0 & 0 & 0 & 0 & 0 & 0 & 1 & 0 & 0 & 1 \\ 0 & 0 & 0 & 1 & 0 & 0 & 0 & 0 & 0 & 0 & 0 & 0 & 1 & 1 & 0 \\ 0 & 0 & 0 & 0 & 1 & 0 & 0 & 0 & 0 & 0 & 0 & 0 & 1 & 0 & 1 \\ 0 & 0 & 0 & 0 & 0 & 1 & 0 & 0 & 0 & 0 & 0 & 0 & 0 & 1 & 1 \\ 0 & 0 & 0 & 0 & 0 & 0 & 1 & 0 & 0 & 0 & 0 & 1 & 1 & 1 & 0 \\ 0 & 0 & 0 & 0 & 0 & 0 & 0 & 1 & 0 & 0 & 0 & 1 & 1 & 0 & 1 \\ 0 & 0 & 0 & 0 & 0 & 0 & 0 & 0 & 1 & 0 & 0 & 1 & 0 & 1 & 1 \\ 0 & 0 & 0 & 0 & 0 & 0 & 0 & 0 & 0 & 1 & 0 & 0 & 1 & 1 & 1 \\ 0 & 0 & 0 & 0 & 0 & 0 & 0 & 0 & 0 & 0 & 1 & 1 & 1 & 1 & 1 \end{pmatrix}$$

6. *The codeword 0110010 was received in a system using a Hamming code with the following generator matrix. What was the transmitted message?*

$$\begin{pmatrix} 1 & 0 & 0 & 0 & 1 & 1 & 0 \\ 0 & 1 & 0 & 0 & 1 & 0 & 1 \\ 0 & 0 & 1 & 0 & 0 & 1 & 1 \\ 0 & 0 & 0 & 1 & 1 & 1 & 1 \end{pmatrix}$$

By considering the generator matrix as $(I_4 \vdots P)$, we can write $H^T = \begin{pmatrix} P \\ I_3 \end{pmatrix}$ which gives

$$\begin{pmatrix} 1 & 1 & 0 \\ 1 & 0 & 1 \\ 0 & 1 & 1 \\ 1 & 1 & 1 \\ 1 & 0 & 0 \\ 0 & 1 & 0 \\ 0 & 0 & 1 \end{pmatrix}.$$

From this, we can calculate the syndrome as 100. This corresponds to the syndrome for an error pattern of 0000100. The transmitted codeword should therefore have been 0110110, and the message was 0110.

7. A binary coding scheme for messages is constructed from two information symbols b_1 and b_2 and three parity symbols (p_1, p_2 and p_3). The generator matrix is as follows:

$$G = \begin{pmatrix} 1 & 0 & 1 & 1 & 1 \\ 0 & 1 & 1 & 0 & 1 \end{pmatrix}$$

(a) How many codewords are there? What are the codewords?
   From the generator matrix, $n = 5$ (number of columns) and $k = 2$ (number of rows). Since this is a binary code and $k = 2$, there are 4 codewords.

(b) What is the minimum distance of the code?
   The codewords are 00000 (message 00), 10111 (message 10), 01101 (message 01) and 11010 (message 11).

(c) What is the theoretical error correcting capability of this code?
   The minimum distance is the minimum weight of a non-zero codeword. The codewords have weight 4, 3 and 3, so the minimum distance is 3. The code can theoretically correct one error or detect up to two errors.

(d) Determine the parity check matrix H.
   The parity check matrix is

$$\begin{pmatrix} 1 & 1 & 1 & 0 & 0 \\ 1 & 0 & 0 & 1 & 0 \\ 1 & 1 & 0 & 0 & 1 \end{pmatrix}.$$

(e) What is the syndrome for the following error patterns: (01000), (00101), (10010), and (11111)?
   The syndrome for error pattern (01000) is 101
   The syndrome for error pattern (00101) is 101
   The syndrome for error pattern (10010) is 101
   The syndrome for error pattern (11111) is 101

(f) A codeword (11010) is generated, which is distorted with the following error pattern: (10010). If the code is used for error correction, what will the decoded codeword be? Explain your choice.
   The codeword (11010) distorted by error (10010) gives a received codeword of (01000). This results in a syndrome of 101. If the code is used for error correction, this syndrome would correspond to a single error in the second bit, so the codeword would be interpreted as 00000 corrupted by 01000. The message would be interpreted as 00.

Note the code makes a mistake, since a codeword of 11010 corresponds to the message 11. This is because it is only single error correcting, and if two errors occur, the codeword is corrupted so severely it is moved within the correcting distance of a different codeword. If the code was used for error correction only, two errors could be dealt with. Indeed, the syndrome is non-zero, so we would have detected the fact that an error occurred. In the error detection case, we would not attempt correction.

8. A binary code is constructed by adding three parity check symbols to three information symbols (denoted by b_1, b_2 and b_3). For these code symbols:

$$
\begin{aligned}
c_1 &= b_1 \\
c_2 &= b_2 \\
c_3 &= b_3 \\
c_4 &= b_2 + b_3 \\
c_5 &= b_1 + b_3 \\
c_6 &= b_1 + b_2
\end{aligned}
$$

(a) What are the codewords?

The codewords can be calculated either directly from the given equations or by constructing the generator matrix. All possible values of b_1 to b_3 give

b_1	b_2	b_3	x_1	x_2	x_3	x_4	x_5	x_6
0	0	0	0	0	0	0	0	0
0	0	1	0	0	1	1	1	0
0	1	0	0	1	0	1	0	1
0	1	1	0	1	1	0	1	1
1	0	0	1	0	0	0	1	1
1	0	1	1	0	1	1	0	1
1	1	0	1	1	0	1	1	0
1	1	1	1	1	1	0	0	0

The codewords are therefore 000000, 001110, 010101, 011011, 100011, 101101, 110110, and 111000.

(b) Determine the generator matrix G and the parity check matrix H.

G can be constructed directly since we are looking for the various coefficients g_{ij} in the following.

$$
\begin{pmatrix} m_1 & m_2 & m_3 \end{pmatrix}
\begin{pmatrix}
g_{11} & g_{12} & g_{13} & g_{14} & g_{15} & g_{16} \\
g_{21} & g_{22} & g_{23} & g_{24} & g_{25} & g_{26} \\
g_{31} & g_{32} & g_{33} & g_{34} & g_{35} & g_{36}
\end{pmatrix}
= \begin{pmatrix} c_1 & c_2 & c_3 & c_4 & c_5 & c_6 \end{pmatrix}
$$

For example, $c_1 = m_1 g_{11} + m_2 g_{21} + m_3 g_{31}$. Substituting b for the message bits m and x for the codeword bits c yields that

$$
G = \begin{pmatrix}
1 & 0 & 0 & 0 & 1 & 1 \\
0 & 1 & 0 & 1 & 0 & 1 \\
0 & 0 & 1 & 1 & 1 & 0
\end{pmatrix}
$$

The parity check matrix is therefore

$$
H = \begin{pmatrix}
0 & 1 & 1 & 1 & 0 & 0 \\
1 & 0 & 1 & 0 & 1 & 0 \\
1 & 1 & 0 & 0 & 0 & 1
\end{pmatrix}
$$

(c) *What is the minimum Hamming distance of the code? What is its error correction and detection capability?*
The Hamming distance is the minimum weight of the non-zero codewords, which is 3. The code can correct one error or detect up to two errors.

9. *A binary linear block code is constructed from a single parity check (SPC) code where for every three information symbols one parity check symbol is added so that there is an even number of 1s in each codeword.*

(a) *Determine the G and H matrices for this code.*
The code has three information symbols. Let these be denoted m_1 m_2 and m_3. A parity bit, p, is added such that $p = 1$ if $m_1 + m_2 + m_3 = 1$, or $p = 0$ if $m_1 + m_2 + m_3 = 0$. Note that we are working with single binary digits, so $0 + 0 = 0, 0 + 1 = 1, 1 + 0 = 1,$ $1 + 1 = 0$. Therefore, we have
$p = m_1 + m_2 + m_3$. The codeword is $c_1 \ c_2 \ c_3 \ c_4$ with $c_1 = m_1, c_2 = m_2, c_3 = m_3,$ $c_4 = p = m_1 + m_2 + m_3$. This gives

$$G = \begin{pmatrix} 1 & 0 & 0 & 1 \\ 0 & 1 & 0 & 1 \\ 0 & 0 & 1 & 1 \end{pmatrix}$$

$$H = \begin{pmatrix} 1 & 1 & 1 & 1 \end{pmatrix}$$

(b) *What is its minimum distance?*
The minimum distance is 2, since the codewords are 0000, 0011, 0101, 0110, 1001, 1010, 1100, 1111.
(There are two possible syndromes: 0, for no error, or 1, for a single bit error. Note that any two, or even number of errors will also give a syndrome of 0, and so will go undetected.)

10. *A binary linear code has the following parity check matrix.*

$$H = \begin{pmatrix} 1 & 1 & 1 & 0 & 1 & 0 & 0 \\ 0 & 0 & 1 & 1 & 1 & 1 & 0 \\ 1 & 0 & 0 & 0 & 1 & 1 & 1 \end{pmatrix}$$

(a) *What are n and k for this code? How many codewords does the code have?*
The H matrix is a 3×7 matrix, so the G matrix will be 4×7. $n = 7$ and $k = 4$, and there are 16 codewords.

(b) *By expressing the H matrix as $[P^\top I]$, where I is the identity matrix, give a systematic generator matrix (i.e. in the form $[IP]$, so the first k bits of the codeword are the same as the message).*
Rearranging the rows of H (row 2 = row 1 + row 2, row 3 = row 1 + row 2 + row 3) gives

$$H = \begin{pmatrix} 1 & 1 & 1 & 0 & 1 & 0 & 0 \\ 1 & 1 & 0 & 1 & 0 & 1 & 0 \\ 1 & 0 & 1 & 1 & 0 & 0 & 1 \end{pmatrix}$$

This means that G is

$$G = \begin{pmatrix} 1 & 0 & 0 & 0 & 1 & 1 & 1 \\ 0 & 1 & 0 & 0 & 1 & 1 & 0 \\ 0 & 0 & 1 & 0 & 1 & 0 & 1 \\ 0 & 0 & 0 & 1 & 0 & 1 & 1 \end{pmatrix}$$

(c) *List all the codewords.*

Message	Codeword	Message	Codeword	Message	Codeword	Message	Codeword
0000	0000000	0100	0100110	1000	1000111	1100	1100001
0001	0001011	0101	0101101	1001	1001100	1101	1101010
0010	0010101	0110	0110011	1010	1010010	1110	1110100
0011	0011110	0111	0111000	1011	1011001	1111	1111111

(d) *What is the minimum distance of the code? How many errors can it correct?*
The minimum distance is 3. It can correct 1 error.

(e) *After transmitting two codewords through a binary symmetric channel using the systematic form of the code, (1000001) and (1001100) are received. Calculate the syndrome for each of these received codewords. Determine the most probable transmitted codeword in both cases, and from this the most likely message.*
In the first case the syndrome is (011), suggesting an error in the second bit, and the codeword (1100001), giving a message of 1100. In the second case, the syndrome is (000), so the codeword would seem to be correct and the message is 1001.

(f) *In this example, assume that the binary symmetric channel has a transition probability of < 0.5. What does this say about the relative probabilities of errored and error free bits? If the transition probability was known to be more than 0.5, is there a simple way of using codes like this one, and if so, how?*
If the transition probability is < 0.5, errors are less likely to occur than uncorrupted bits, and the nearest codeword is therefore the most likely one. If the transition probability is > 0.5, then all the received bits can be inverted, which would result in a transition probability less than 0.5, so the assumption that the nearest codeword was the correct one could be used again. If the transition probability is 0.5, then there is no code which can be used to send information and the channel capacity is zero, as will be shown in the next chapter.

8.4.2   Answers to Questions on Convolutional Codes

1. *A convolutional encoder has the following encoding circuit.*

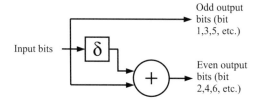

(a) *Draw the code trellis for this code.*

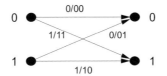

(b) *What is the constraint length of the code? For this code, the minimum non-zero path is generated from the input 100... Find the minimum distance for the code, and state its error correcting capability.*

The constraint length is 2. 1000...produces 110100000...which has a weight of 3. The code can therefore correct 1 errors occurring within a window of 4 bits.

(c) *By repeating the trellis diagram for each input bit, and clearly showing each change in state, encode the sequence 1 0 1 1 0 1 0 0.*

	1	2	3	4	5	6	7	8
Input	1	0	1	1	0	1	0	0
Current State	0	1	0	1	1	0	1	0
Next State	1	0	1	1	0	1	0	0
Output	11	01	11	10	01	11	01	00

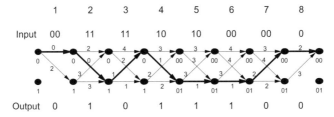

(d) *By drawing the trellis diagram showing the path weights at each step, use the Viterbi algorithm to perform maximum likelihood decoding on 00 11 11 11 10 10 00 00. Hence state the most likely transmitted sequence.*

	1	2	3	4	5	6	7	8
Input	00	11	11	10	10	00	00	0

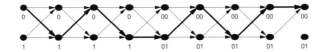

| Output | 0 | 1 | 0 | 1 | 1 | 1 | 0 | 0 |

2. *A convolutional encoder has the following encoding circuit.*

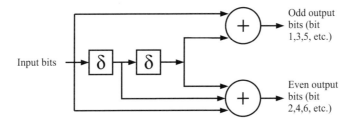

(a) *Draw the code trellis for this code.*

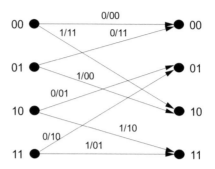

(b) *What is the constraint length of the code? For this code, the minimum non-zero path is generated from the input 100... Find the minimum distance for the code, and state its error correcting capability.*

The constraint length is 3. 1000...produces 110111000... which has a weight of 5. The code can therefore correct 2 errors occurring within a window of 6 bits.

(c) *By repeating the trellis diagram for each input bit, and clearly showing each change in state, encode the sequence 1 0 1 1 0 1 0 0.*

	1	2	3	4	5	6	7	8
Input	1	0	1	1	0	1	0	0
Current State	00	10	01	10	11	01	10	01
Next State	10	01	10	11	01	10	01	00
Output	11	01	00	01	11	11	01	00

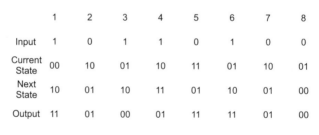

(d) *By drawing the trellis diagram showing the path weights at each step, use the Viterbi algorithm to perform maximum likelihood decoding on 00 01 10 01 11. Hence state the most likely transmitted sequence.*

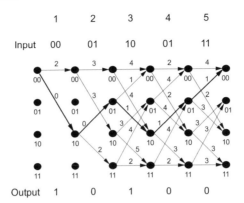

8.4.2.1 Answers to Questions on LANs and MANs

1. *A LAN based upon IEEE 802.3 CSMA/CD 10Base2 is used to carry data between two computers located/cabled 15m apart.*

   (a) *If a simple connectionless service is employed by LLC, how long will it take to transfer a 1 Mbyte file between these devices if no other devices are using the network? Assume a 9.6 μs inter-frame gap and maximum frame size.*
   Payload = 1500 bytes and file size =1 Mbyte (2^{20} bytes), so 700 frames will need to be transmitted.
   Each frame takes $1518 \times 8 \times 0.1\mu s$ to transmit = 1.2144 ms plus interframe gap of 9.6 μs and preamble delay of 5.6 μs. Total delay is 1.2296 ms. Propagation delay can be included – many options to choose: $\frac{1}{2}$ of collision window time (25.6μs), max segment propagation delay (12.5 μs) or propagation delay between device 15 metres apart (0.075 μs). This is small compared to transmission delay so we will neglect it.
   Time to transmit file= 700×1.2296ms = 860.72 ms.

   (b) *How long will the file transfer take if the LLC uses an acknowledged datagram service?*
   For acknowledged service, an ack frame is sent in response to frame being transmitted successfully – assume stop and wait and no errors so that delay per frame is time to transmit frame and ack. The latter is a minimum length frame that incurs transmission delay of 51.2 μS. Allow for preamble and interframe gap, gives extra delay of 47.04 ms for all 700 frames and a total delay of $860.72 + 47.04 = 907.76$ ms.

   (c) *If other devices are now using the network such that each frame is subject to an average of 4 collisions, estimate the time taken to transfer the 1 Mbyte file using the unacknowledged LLC service.*
   Each frame has to be transmitted 5 times (4 collisions plus final success). Also included is back-off time for each collision. 1st collision results in mean delay of $\frac{1}{2}$ unit, 2nd collision results in mean delay of 1.5 units, 3rd collision results in mean delay of 3.5 units while 4th collision results in mean delay of 7.5 units giving a total delay of 13 units. (strictly not correct to add the back-off delays in this manner but is simplest method) Back-off delay per frame= $13 \times 51.2\mu s$= 0.6656 mS. Total back-off delay is 700×0.6656ms = 466 ms. The total transmission delay = $5 \times 860.72 + 466$ms = 4.7 seconds. This is only an estimate and perhaps is lower than would really be experienced under such noisy conditions.

2. *Estimate the time taken to transmit 1 Mbyte of data using full duplex transmission over High Speed and Gigabit Ethernet for frame sizes of 248 and 1500 octet payloads.*

   - 100Base – payload of 248 octets – frame of 266 octets, transmission delay of $8 \times 266 \times 0.01 = 21.28\mu s$
   - 100Base – payload of 1500 octets – frame of 1518 octets, transmission delay of $8 \times 1518 \times 0.01 = 121.4\mu s$
   - 1000Base – payload of 248 octets – frame of 266 octets, transmission delay of $8 \times 266 \times 0.001 = 2.128\mu s$ but need to allow for collision slot.
   - 1000Base – payload of 1500 octets – frame of 1518 octets, transmission delay of $8 \times 1518 \times 0.001 = 12.14\mu s$

3. *Determine the efficiency of the transmission process associated with half duplex Gigabit Ethernet when transmitting minimum size frames, with and without frame-bursting.*
   Without frame busting, efficiency is $64/512 = 12.5\%$ or $46/512 = 8.98\%$.
   With frame bursting, one can transmit 1 frame within the collision slot and 107 frames (max) during the remaining 8192–512 bytes of the frame burst. The first frame is only 8.98% efficient while the remaining 100 or so are $46/72 = 0.634$ efficient. Overall efficiency is therefore just lower than 63%.

4. *For both 100Base and 1000Base CSMA/CD networks, estimate how long it will take for 1 Mbyte of data to be transferred via a contention-free half duplex configuration when frame payloads are 248 and 1500 octets.*
   1500 octet payloads imply 700 frames to be transmitted. Overall delay for both 100Base and 1000Base options is simply 700 times the individual frame delay, i.e. 85 or 8.5 ms respectively.
   248 octet payloads imply 4229 frames to be sent. For 100Base the delay is $4229 \times 21.28 = 90$ ms. For 1000Base we must include frame bursting – one burst of 8192 octets is equivalent to 29 frames. Thus 145.9 bursts are required, given a total delay of $65.536\mu s \times 145.9 = 9.56$ ms.

5. *A MAN based upon the FDDI standard consists of 50 nodes, 4km apart, connected by a 100 Mbit/s ring. Each node generates an average of 100 kbit/s data traffic, while half of the nodes generates 256 kbit/s of voice traffic and the other half generates 2.5 Mbit/s of video traffic which has a real-time delay constraint such that successive frames must not take longer than 12 milli-seconds to traverse the MAN.*

   (a) *If it takes the MAC token 600 micro-seconds to circulate the network under no-load connections, estimate the actual and maximum possible utilisation for this network under these conditions.*
   Actual Utilisation = Actual Load/Total Capacity = $[(50 \times 0.1) + (25 \times 0.256) + (25 \times 2.5)]/100 = 73.9\%$, where all numbers are in Mbit/s.
   In FDDI, max. end-to-end delay is given by $2 \times$ TTRT, which is given by 12 ms.
   Thus, TTRT = 6 ms
   Ring Latency (RL) = 0.6 ms from the question. Number of Stations (N) = 50.
   Max Utilisation $= \frac{N(\text{TTRT} - \text{RL})}{(N \times \text{TTRT}) + \text{RL}} = 89.9\%$ or
   Max Utilisation $= \frac{\text{TTRT} - \text{RL}}{\text{TTRT}} = 90\%$, if N is large.

   (b) *What options are available to improve the maximum efficiency and what are the consequences these will have on overall performance?*
   To improve efficiency, we can try to reduce RL or increase TTRT.

The latter, will increase end-to-end delay so will impact upon the ability to support real-time services. The flexibility to change this parameter depends upon services supported. RL is a physical parameter and can be reduced by reducing the size of the network. TTRT is the dominant parameter.

8.5 Answers to Questions on the Channel Perspective

8.5.1 Answers to Questions on Channel Capacity

1. *A binary symmetric channel with a transition probability of $\frac{1}{4}$ is cascaded with a second channel also with a transition probability of $\frac{1}{4}$ so that the output of the first channel feeds the second channel. Denote the input to the first channel to be X, the output of the first channel and the input to the second to be Y, and the output of the second to be Z. The source at X has a probability of a 0 as $\frac{1}{3}$, and the probability of a 1 as $\frac{2}{3}$.*

(a) *Calculate $H(X|Y)$*

$H(X|Y) = \Sigma_k \left(H(X \mid Y = y_k)\right) p(y_k)$. To simplify calculations, we first calculate all the relevant probabilities. The channel matrix is $\begin{bmatrix} \frac{3}{4} & \frac{1}{4} \\ \frac{1}{4} & \frac{3}{4} \end{bmatrix}$.

$p(X=0) = \frac{1}{3}$	$p(Y=0|X=0) = \frac{3}{4}$	$p(Y=0,X=0) = \frac{1}{4}$	$p(Y=0) = \frac{5}{12}$	$p(X=0|Y=0) = \frac{3}{5}$
	$p(Y=1|X=0) = \frac{1}{4}$	$p(Y=1,X=0) = \frac{1}{12}$		$p(X=1|Y=0) = \frac{2}{5}$
$p(X=1) = \frac{2}{3}$	$p(Y=0|X=1) = \frac{1}{4}$	$p(Y=0,X=1) = \frac{1}{6}$	$p(Y=1) = \frac{7}{12}$	$p(X=0|Y=1) = \frac{1}{7}$
	$p(Y=1|X=1) = \frac{3}{4}$	$p(Y=1,X=1) = \frac{1}{2}$		$p(X=1|Y=1) = \frac{6}{7}$

(Note that it is quite possible to have conditional probabilities out of time order, so even though event X happens before event Y, we can still define $p(X=0|Y=0)$, for example, as the probability the event X had been 0 given that event Y was seen to be 0.)

$$
\begin{aligned}
H(X|Y=0) &= -p(X=0|Y=0)\log_2(p(X=0|Y=0)) - p(X=1|Y=0)\log_2(p(X=1|Y=0)) \\
&= -3/5\log_2 3/5 - 2/5\log_2 2/5 = 0.97095 \text{ bits} \\
H(X|Y=1) &= -p(X=0|Y=1)\log_2(p(X=0|Y=1)) - p(X=1|Y=1)\log_2(p(X=1|Y=1)) \\
&= -1/7\log_2 1/7 - 6/7\log_2 6/7 = -0.59167 \text{ bits} \\
H(X|Y) &= H(X|Y=0)p(Y=0) + H(X|Y=1)p(Y=1) \\
&= 0.97095 \times \frac{5}{12} + 0.59167 \times \frac{7}{12} = 0.40456 + 0.34514 = 0.74970 \text{ bits}
\end{aligned}
$$

(b) *Calculate the mutual information between X and Y.*

The mutual information between X and Y is $I(X;Y) = H(X) - H(X|Y)$
$= -\frac{1}{3}\log_2\frac{1}{3} - \frac{2}{3}\log_2\frac{2}{3} - 0.74970 = 0.9183 - 0.7497 = 0.169$ bits

(c) *Calculate $H(X|Z)$.*

(d) *Calculate the mutual information between X and Z.*

The same techniques as above can be applied.

2. *A Binary Erasure Channel has the following form. What is its capacity in terms of p?*

The transition matrix is $\begin{bmatrix} p & 1-p & 0 \\ 0 & 1-p & p \end{bmatrix}$. Let the input be X and the output be Y. Let the probability of a 0 at X be a. Channel capacity is the maximum value of mutual information. $I(X;Y) = H(Y) - H(Y|X)$.

The relevant probabilities are

p(X=0) = a	p(Y=0|X=0) = p	p(Y=0,X=0) = pa	p(Y=0) = pa
	p(Y=U|X=0) = 1–p	p(Y=U,X=0) = (1–p)a	
	p(Y=1|X=0) = 0	p(Y=1,X=0) = 0	p(Y=U) = (1–p)
p(X=1) = 1 – a	p(Y=0|X=1) = 0	p(Y=0,X=1) = 0	
	p(Y=U|X=1) = 1–p	p(Y=U,X=1) = (1–p)(1 – a)	p(Y=1) = p(1 – a)
	p(Y=1|X=1) = p	p(Y=1,X=1) = p(1 – a)	

$$
\begin{aligned}
H(Y) \ =\ & p(Y{=}0)\log_2(p(Y{=}0)) - p(Y{=}U)\log_2(p(Y{=}U)) - p(Y{=}1)\log_2(p(Y{=}1)) \\
=\ & -pa\log_2 pa - (1-p)\log_2(1-p) - p(1-a)\log_2 p(1-a) \\
=\ & -p\log_2 p - (1-p)\log_2(1-p) - pa\log_2 a - p(1-a)\log_2(1-a)
\end{aligned}
$$

$H(Y|X) = p(X{=}0)H(Y|X{=}0) + p(X{=}1)H(Y|X{=}1) = a\left(-p\log_2 p - (1-p)\log_2(1-p)\right)$
$+(1-a)(-(1-p)\log_2(1-p) - p\log_2 p) = -p\log_2 p - (1-p)\log_2(1-p)$
$H(Y) - H(Y|X) = -p\log_2 p - (1-p)\log_2(1-p) - pa\log_2 a - p(1-a)\log_2(1-a) + p\log_2 p +$
$(1-p)\log_2(1-p) = -pa\log_2 a - p(1-a)\log_2(1-a) = p(a\log_2 a - (1-a)\log_2(1-a))$
Now $(a\log_2 a - (1-a)\log_2(1-a))$ is the entropy of the binary source X. Its maximum
value is 1, and so the maximum capacity of the channel is p.

3. *A telephone channel has a bandwidth of 3.4kHz. Calculate (a) the channel capacity if the signal to noise ratio is 30dB and (b) the minimum signal to noise ratio required (theoretically) to sustain an information transmission of 4800 bits per second.*

   (a) Using Shannon's Law, $C = B\log_2\left(\frac{S+N}{N}\right)$ or $C = B\log_2\left(1 + \frac{S}{N}\right)$. The signal to noise ratio is given as 30dB. In linear terms, this is $10^3 = 1000$. Therefore, $\frac{S}{N} = 1000$
   $B = 3400$, so $C = 3400\log_2(1 + 1000) = 3400 \times 9.96722 = 33888$ bits/sec

   (b) In order to sustain 4800 bits/s, $B\log_2\left(1 + \frac{S}{N}\right) \geq 4800$. $B = 3400$, so $\log_2\left(1 + \frac{S}{N}\right) \geq \frac{4800}{3400}$, $1 + \frac{S}{N} \geq 2^{1.412}$, $\frac{S}{N} \geq 2.66$, or 4.25dB

Index